REFLECTIONS ON THE DECLINE OF SCIENCE IN ENGLAND,

AND ON SOME OF ITS CAUSES.

By Charles Babbage

Reedited by Philippe ballin

In February 2017

DEDICATION.

HAD I INTENDED TO DEDICATE THIS VOLUME, I SHOULD HAVE INSCRIBED IT TO A NOBLEMAN WHOSE EXERTIONS IN PROMOTING EVERY OBJECT THAT CAN ADVANCE SCIENCE REFLECT LUSTRE UPON HIS RANK. BUT THE KINDNESS OF HIS NATURE MIGHT HAVE BEEN PAINED AT HAVING HIS NAME CONNECTED WITH STRICTURES, PERHAPS TOO SEVERELY JUST. I SHALL, THEREFORE, ABSTAIN FROM MENTIONING THE NAME OF ONE WHO WILL FEEL THAT HE HAS COMMANDED MY ESTEEM AND RESPECT.

C. BABBAGE.

DORSET STREET, MANCHESTER SQUARE, 29th April, 1830.

PREFACE.

Of the causes which have induced me to print this volume I have little to say; my own opinion is, that it will ultimately do some service to science, and without that belief I would not have undertaken so thankless a task. That it is too true not to make enemies, is an opinion in which I concur with several of my friends, although I should hope that what I have written will not give just reason for the permanence of such feelings. On one point I shall speak decidedly, it is not connected in any degree with the calculating machine on which I have been engaged; the causes which have led to it have been long operating, and would have produced this result whether I had ever speculated on that subject, and whatever might have been the fate of my speculations.

If any one shall endeavour to account for the opinions stated in these pages by ascribing them to any imagined circumstance peculiar to myself, I think he will be mistaken. That science has long been neglected and declining in England, is not an opinion originating with me, but is shared by many, and has been expressed by higher authority than mine. I shall offer a few notices on this subject, which, from their scattered position, are unlikely to have met the reader's attention, and which, when combined with the facts I have detailed in subsequent pages, will be admitted to deserve considerable attention. The following extract from the article Chemistry, in the Encyclopaedia Metropolitana, is from the pen of a gentleman equally qualified by his extensive reading, and from his acquaintance with foreign nations, to form an opinion entitled to respect. Differing from him widely as to the cause, I may be permitted to cite him as high authority for the fact.

"In concluding this most circumscribed outline of the History of Chemistry, we may perhaps be allowed to express a faint shade of regret, which, nevertheless, has frequently passed over our minds within the space of the last five or six years. Admiring, as we most sincerely do, the electro-magnetic discoveries of Professor Oersted and his followers, we still, as chemists, fear that our science has suffered some degree of neglect in consequence of them. At least, we remark that, during this period, good chemical analyses and researches have been rare in England; and yet, it must be confessed, there is an ample field for chemical discovery. How scanty is our knowledge of the suspected fluorine! Are we sure that we understand the nature of nitrogen? And yet these are amongst our elements. Much has been done by Wollaston, Berzelius, Guy-Lussac, Thenard, Thomson, Prout, and others, with regard to the doctrine of definite proportions; but there yet remains the Atomic Theory. Is it a representation of the laws of nature, or is it not?"—-CHEMISTRY, ENCYC. METROP. p.596.

When the present volume was considerably

advanced, the public were informed that the late Sir Humphry Davy had commenced a work, having the same title as the present, and that his sentiments were expressed in the language of feeling and of eloquence. It is to be hoped that it may be allowed by his friends to convey his opinions to posterity, and that the writings of the philosopher may enable his contemporaries to forget some of the deeds of the President of the Royal Society.

Whatever may be the fate of that highly interesting document, we may infer his opinions upon this subject from a sentiment expressed in his last work:—

"—But we may in vain search the aristocracy now for philosophers."——"There are very few persons who pursue science with true dignity; it is followed more as connected with objects of profit than those of fame."—SIR H. DAVY'S CONSOLATIONS IN TRAVEL.

The last authority which I shall adduce is more valuable, from the varied acquirements of its author, and from the greater detail into which he enters. "We have drawn largely, both in the present Essay, and in our article on LIGHT, from the ANNALES DE CHEMIE, and we take this ONLY opportunity distinctly to acknowledge our obligations to that most admirably conducted work. Unlike the crude and undigested scientific matter which suffices, (we are ashamed to say it) for the monthly and quarterly amusement of our own countrymen, whatever is admitted into ITS pages, has at least been taken pains with, and, with few exceptions, has sterling merit. Indeed, among the original communications which abound in it, there are few which would misbecome the first academical collections; and if any thing could diminish our regret at the long suppression of those noble memoirs, which are destined to adorn future volumes of that of the Institute, it would be the masterly abstracts of them which from time to time appear in the ANNALES, either from the hands of the authors, or from the reports rendered by the committees appointed to examine them; which latter, indeed, are universally models of their kind, and have contributed, perhaps more than any thing, to the high scientific tone of the French SAVANS. What author, indeed, but will write his best, when he knows that his work, if it have merit, will immediately be reported on by a committee, who will enter into all its meaning; understand it, however profound: and, not content with MERELY understanding it, pursue the trains of thought to which it leads; place its discoveries and principles in new and unexpected lights; and bring the whole of their knowledge of collateral subjects to bear upon it. Nor ought we to omit our acknowledgement to the very valuable Journals of Poggendorff and Schweigger. Less exclusively national than their Gallic compeer, they present a picture of the actual progress of physical science throughout Europe. Indeed, we have been often astonished to see with what celerity every thing, even moderately valuable in the scientific publications of this country, finds its way into their pages. This ought to encourage our men of science. They have a larger audience, and a wider sympathy than they are perhaps aware of; and however disheartening the general diffusion of smatterings of a number of subjects, and the almost equally general indifference to profound knowledge in any, among their own countrymen, may be, they may rest assured that not a fact they may discover, nor a good experiment they may make, but is instantly repeated, verified, and commented upon, in Germany, and, we may add too, in Italy. We wish the obligation were mutual. Here, whole branches of continental discovery are unstudied, and indeed almost unknown, even by name. It is in vain to conceal the melancholy truth. We are fast dropping behind. In mathematics we have long since drawn the rein, and given over a hopeless race. In chemistry the case is not much better. Who can tell us any thing of the Sulfo-salts? Who will explain to us the laws of Isomorphism? Nay, who among us has even verified Thenard's experiments on the oxygenated acids,—Oersted's and Berzelius's on the radicals of the earths,—Balard's and Serrulas's on the combinations of Brome,—and a hundred other splendid trains of research in that fascinating science? Nor need we stop here. There are, indeed, few sciences which would not furnish matter for similar remark. The causes are at once obvious and deep-seated; but this is not the place to discuss them."—MR. HERSCHEL'S TREATISE ON SOUND, printed in the ENCYCLOPAEDIA METROPOLITANA.

With such authorities, I need not apprehend much doubt as to the fact of the decline of science in England: how far I may have pointed out some of its causes, must be left to others to decide.

Many attacks have lately been made on the

conduct of various scientific bodies, and of their officers, and severe criticism has been lavished upon some of their productions. Newspapers, Magazines, Reviews, and Pamphlets, have all been put in requisition for the purpose. Odium has been cast upon some of these for being anonymous. If a fact is to be established by testimony, anonymous assertion is of no value; if it can be proved, by evidence to which the public have access, it is of no consequence (for the cause of truth) who produces it. A matter of opinion derives weight from the name which is attached to it; but a chain of reasoning is equally conclusive, whoever may be its author.

Perhaps it would be better for science, that all criticism should be avowed. It would certainly have the effect of rendering it more matured, and less severe; but, on the other hand, it would have the evil of frequently repressing it altogether, because there exists amongst the lower ranks of science, a "GENUS IRRITABILE," who are disposed to argue that every criticism is personal. It is clearly the interest of all who fear inquiries, to push this principle as far as possible, whilst those whose sole object is truth, can have no apprehensions from the severest scrutiny. There are few circumstances which so strongly distinguish the philosopher, as the calmness with which he can reply to criticisms he may think undeservedly severe. I have been led into these reflections, from the circumstance of its having been stated publicly, that I was the author of several of those anonymous writings, which were considered amongst the most severe; and the assertion was the more likely to be credited, from the fact of my having spoken a few words connected with one of those subjects at the last anniversary of the Royal Society. [I merely observed that the agreement made with the British Museum for exchanging the Arundel MSS. for their duplicates, (which had just been stated by the President,) was UNWISE;—because it was not to be expected that many duplicates should be found in a library like that of the Museum, weak in the physical and mathematical sciences: that it was IMPROVIDENT and UNBUSINESSLIKE;— because it neither fixed the TIME when the difference was to be paid, in case their duplicates should be insufficient; nor did it appear that there were any FUNDS out of which the money could be procured: and I added, that it would be more advantageous to sell the MSS., and purchase the books we wanted with the produce.] I had hoped in that diminutive world, the world of science, my character had been sufficiently known to have escaped being the subject of such a mistake; and, in taking this opportunity of correcting it, I will add that, in the present volume, I have thought it more candid to mention distinctly those whose line of conduct I have disapproved, or whose works I have criticised, than to leave to the reader inferences which he might make far more extensive than I have intended. I hope, therefore, that where I have depicted species, no person will be so unkind to others and unjust to me, as to suppose I have described individuals.

With respect to the cry against personality, which has been lately set up to prevent all inquiry into matters of scientific misgovernment, a few words will suffice.

I feel as strongly as any one, not merely the impropriety, but the injustice of introducing private character into such discussions. There is, however, a maxim too well established to need any comment of mine. The public character of every public servant is legitimate subject of discussion, and his fitness or unfitness for office may be fairly canvassed by any person. Those whose too sensitive feelings shrink from such an ordeal, have no right to accept the emoluments of office, for they know that it is the condition to which all must submit who are paid from the public purse.

The same principle is equally applicable to Companies, to Societies, and to Academies. Those from whose pocket the salary is drawn, and by whose appointment the officer was made, have always a right to discuss the merits of their officers, and their modes of exercising the duties they are paid to perform.

This principle is equally applicable to the conduct of a Secretary of State, or to that of a constable; to that of a Secretary of the Royal Society, or of an adviser to the Admiralty.

With respect to honorary officers, the case is in some measure different. But the President of a society, although not recompensed by any pecuniary remuneration, enjoys a station, when the body over which he presides possesses a high character, to which many will aspire, who will esteem themselves amply repaid for the time they devote to the office, by the consequence attached to it in public estimation. He, therefore, is answerable to the

Society for his conduct in their chair.

There are several societies in which the secretaries, and other officers, have very laborious duties, and where they are unaided by a train of clerks, and yet no pecuniary remuneration is given to them. Science is much indebted to such men, by whose quiet and unostentatious labours the routine of its institutions is carried on. It would be unwise, as well as ungrateful, to judge severely of the inadvertencies, or even of the negligence of such persons: nothing but weighty causes should justify such a course.

Whilst, however, I contend for the principle of discussion and inquiry in its widest sense, because I consider it equally the safeguard of our scientific as of our political institutions, I shall use it, I hope, temperately; and having no personal feelings myself, but living in terms of intercourse with almost all, and of intimacy with several of those from whom I most widely differ, I shall not attempt to heap together all the causes of complaint; but, by selecting a few in different departments, endeavour to convince them that some alteration is essentially necessary for the promotion of that very object which we both by such different roads pursue.

I have found it necessary, in the course of this volume, to speak of the departed; for the misgovernment of the Royal Society has not been wholly the result of even the present race. It is said, and I think with justice, in the life of Young, inserted amongst Dr. Johnson's, that the famous maxim, "DE MORTUIS NIL NISI BONUM," "appears to savour more of female weakness than of manly reason." The foibles and the follies of those who are gone, may, without injury to society, repose in oblivion. But, whoever would claim the admiration of mankind for their good actions, must prove his impartiality by fearlessly condemning their evil deeds. Adopt the maxim, and praise to the dead becomes worthless, from its universality; and history, a greater fable than it has been hitherto deemed.

Perhaps I ought to apologize for the large space I have devoted to the Royal Society. Certainly its present state gives it no claim to that attention; and I do it partly from respect for its former services, and partly from the hope that, if such an Institution can be of use to science in the present day, the attention of its members may be excited to take steps for its restoration. Perhaps I may be blamed for having published extracts from the minutes of its proceedings without the permission of its Council. To have asked permission of the present Council would have been useless. I might, however, have given the substance of what I have extracted without the words, and no one could then have reproached me with any infringement of our rules: but there were two objections to that course. In the first place, it is impossible, even for the most candid, in all cases, to convey precisely the same sentiment in different language; and I thought it therefore more fair towards those from whom I differed, as well as to the public, to give the precise words. Again: had it been possible to make so accurate a paraphrase, I should yet have preferred the risk of incurring the reproach of the Royal Society for the offence, to escaping their censure by an evasion. What I have done rests on my own head; and I shrink not from the responsibility attaching to it.

If those, whose mismanagement of that Society I condemn, should accuse me of hostility to the Royal Society; my answer is, that the party which governs it is not the Royal Society; and that I will only admit the justice of the accusation, when the whole body, becoming acquainted with the system I have exposed, shall, by ratifying it with their approbation, appropriate it to themselves: an event of which I need scarcely add I have not the slightest anticipation.

CONTENTS

DEDICATION.
PREFACE.
REFLECTIONS ON THE DECLINE OF SCIENCE IN ENGLAND
INTRODUCTORY REMARKS.

CHAPTER I. ON THE RECIPROCAL INFLUENCE OF SCIENCE AND EDUCATION.
CHAPTER II. OF THE INDUCEMENTS TO INDIVIDUALS TO CULTIVATE SCIENCE.
SECTION 1. PROFESSIONAL IMPULSES.

SECTION 2. OF NATIONAL ENCOURAGEMENT.

SECTION 3. Of Encouragement from Learned Societies.

CHAPTER III. GENERAL STATE OF LEARNED SOCIETIES IN ENGLAND.

CHAPTER IV. STATE OF THE ROYAL SOCIETY IN PARTICULAR.

SECTION 1. MODE OF BECOMING A FELLOW OF THE ROYAL SOCIETY.

SECTION 2. OF THE PRESIDENCY AND VICE-PRESIDENCIES.

SECTION 3. OF THE SECRETARISHIPS.

SECTION 4. OF THE SCIENTIFIC ADVISERS.

SECTION 5. OF THE UNION OF SEVERAL OFFICES IN ONE PERSON.

SECTION 6. OF THE FUNDS OF THE SOCIETY.

SECTION 7. OF THE ROYAL MEDALS.

SECTION 8. OF THE COPLEY MEDALS.

SECTION 9. OF THE FAIRCHILD LECTURE.

SECTION 10. OF THE CROONIAN LECTURE.

SECTION 11. OF THE CAUSES OF THE PRESENT STATE OF THE ROYAL SOCIETY.

SECTION 12. OF THE PLAN FOR REFORMING THE SOCIETY.

CHAPTER V. OF OBSERVATIONS.

SECTION 1. OF MINUTE PRECISION.

SECTION 2. ON THE ART OF OBSERVING.

SECTION 3. ON THE FRAUDS OF

OBSERVERS.

CHAPTER VI. SUGGESTIONS FOR THE ADVANCEMENT OF SCIENCE IN ENGLAND.

SECTION 1. OF THE NECESSITY THAT MEMBERS OF THE ROYAL SOCIETY SHOULD

SECTION 2. OF BIENNIAL PRESIDENTS.

SECTION 3. OF THE INFLUENCE OF THE COLLEGES OF PHYSICIANS AND SURGEONS

SECTION 4. OF THE INFLUENCE OF THE ROYAL INSTITUTION ON THE ROYAL SOCIETY.

SECTION 5. OF THE TRANSACTIONS OF THE ROYAL SOCIETY.

SECTION 6. ORDER OF MERIT.

SECTION 7. OF THE UNION OF SCIENTIFIC SOCIETIES.

CONCLUSION.

APPENDIX, No. 1.

APPENDIX, No. 2.

APPENDIX, No. 3,

LIST OF THE MEMBERS OF THE ROYAL SOCIETY

REFLECTIONS ON THE DECLINE OF SCIENCE IN ENGLAND, AND ON SOME OF ITS CAUSES.

INTRODUCTORY REMARKS.

It cannot have escaped the attention of those, whose acquirements enable them to judge, and who have had opportunities of examining the state of science in other countries, that in England, particularly with respect to the more difficult and abstract sciences, we are much below other nations, not merely of equal rank, but below several even of inferior power. That a country, eminently distinguished for its mechanical and manufacturing ingenuity, should be indifferent to the progress of inquiries which form the highest departments of that knowledge on whose more elementary truths its wealth and rank depend, is a fact which is well deserving the attention of those who shall inquire into the causes that influence the progress of nations.

To trace the gradual decline of mathematical, and with it of the highest departments of physical science, from the days of Newton to the present, must be left to the historian. It is not within the province of one who, having mixed sufficiently with scientific society in England to see and regret the weakness of some of its greatest ornaments, and to see through and deplore the conduct of its pretended friends, offers these remarks, with the hope that they may excite discussion,—with the conviction that discussion is the firmest ally of truth,—and with the confidence that nothing but the full expression of public opinion can remove the evils that chill the enthusiasm, and cramp the energies of the science of England.

The causes which have produced, and some of the effects which have resulted from, the present state of science in England, are so mixed, that it is difficult to distinguish accurately between them. I shall, therefore, in this volume, not attempt any minute discrimination, but rather present the result of my reflections on the concomitant circumstances which have attended the decay, and at the conclusion of it, shall examine some of the suggestions which have been offered for the advancement of British science.

CHAPTER I. ON THE RECIPROCAL INFLUENCE OF SCIENCE AND EDUCATION.

That the state of knowledge in any country will exert a directive influence on the general system of instruction adopted in it, is a principle too obvious to require investigation. And it is equally certain that the tastes and pursuits of our manhood will bear on them the traces of the earlier impressions of our education. It is therefore not unreasonable to suppose that some portion of the neglect of science in England, may be attributed to the system of education we pursue. A young man passes from our public schools to the universities, ignorant almost of the elements of every branch of useful knowledge; and at these latter establishments, formed originally for instructing those who are intended for the clerical profession, classical and mathematical pursuits are nearly the sole objects proposed to the student's ambition.

Much has been done at one of our universities during the last fifteen years, to improve the system of study; and I am confident that there is no one connected with that body, who will not do me the justice to believe that, whatever suggestions I may venture to offer, are prompted by the warmest feelings for the honour and the increasing prosperity of its institutions. The ties which connect me with

Cambridge are indeed of no ordinary kind.

Taking it then for granted that our system of academical education ought to be adapted to nearly the whole of the aristocracy of the country, I am inclined to believe that whilst the modifications I should propose would not be great innovations on the spirit of our institutions, they would contribute materially to that important object.

It will be readily admitted, that a degree conferred by an university, ought to be a pledge to the public that he who holds it possesses a certain quantity of knowledge. The progress of society has rendered knowledge far more various in its kinds than it used to be; and to meet this variety in the tastes and inclinations of those who come to us for instruction, we have, besides the regular lectures to which all must attend, other sources of information from whence the students may acquire sound and varied knowledge in the numerous lectures on chemistry, geology, botany, history, &c. It is at present a matter of option with the student, which, and how many of these courses he shall attend, and such it should still remain. All that it would be necessary to add would be, that previously to taking his degree, each person should be examined by those Professors, whose lectures he had attended. The pupils should then be arranged in two classes, according to their merits, and the names included in these classes should be printed. I would then propose that no young man, except his name was found amongst the "List of Honours," should be allowed to take his degree, unless he had been placed in the first class of some one at least of the courses given by the professors. But it should still be imperative upon the student to possess such mathematical knowledge as we usually require. If he had attained the first rank in several of these examinations, it is obvious that we should run no hazard in a little relaxing: the strictness of his mathematical trial.

If it should be thought preferable, the sciences might be grouped, and the following subjects be taken together:—

Modern History.
Laws of England.
Civil Law.

Political Economy.
Applications of Science to Arts and Manufactures.

Chemistry.
Mineralogy.
Geology.

Zoology, including Physiology and Comparative Anatomy.
Botany, including Vegetable Physiology and Anatomy.

One of the great advantages of such a system would be, that no young person would have an excuse for not studying, by stating, as is most frequently done, that the only pursuits followed at Cambridge, classics and mathematics, are not adapted either to his taste, or to the wants of his after life. His friends and relatives would then reasonably expect every student to have acquired distinction in

SOME pursuit. If it should be feared that this plan would lead to too great a diversity of pursuits in the same individual, a limitation might be placed upon the number of examinations into which the same person might be permitted to enter. It might also be desirable not to restrict the whole of these examinations to the third year, but to allow the student to enter on some portion of them in the first or second year, if he should prefer it.

By such an arrangement, which would scarcely interfere seriously with our other examinations, we should, I think, be enabled effectually to keep pace with the wants of society, and retaining fully our power and our right to direct the studies of those who are intended for the church, as well as of those who aspire to the various offices connected with our academical institutions; we should, at the same time, open a field of honourable ambition to multitudes, who, from the exclusive nature of our present studies, leave us with but a very limited addition to their stock of knowledge.

Much more might be said on a subject so important to the interests of the country, as well as of our university, but my wish is merely to open it for our own consideration and discussion. We have already done so much for the improvement of our system of instruction, that public opinion will not reproach us for any unwillingness to alter. It is our first duty to be well satisfied that we can improve: such alterations ought only to be the result of a most mature consideration, and of a free interchange of sentiments on the subject, in order that we may condense upon the question the accumulated judgment of many minds.

It is in some measure to be attributed to the defects of our system of education, that scientific knowledge scarcely exists amongst the higher classes of society. The discussions in the Houses of Lords or of Commons, which arise on the occurrence of any subjects connected with science, sufficiently prove this fact, which, if I had consulted the extremely limited nature of my personal experience, I should, perhaps, have doubted.

CHAPTER II. OF THE INDUCEMENTS TO INDIVIDUALS TO CULTIVATE SCIENCE.

Interest or inclination form the primary and ruling motives in this matter: and both these exert greater or less proportionate influence in each of the respective cases to be examined.

SECTION 1. PROFESSIONAL IMPULSES.

A large portion of those who are impelled by ambition or necessity to advance themselves in the world, make choice of some profession in which they imagine their talents likely to be rewarded with success; and there are peculiar advantages resulting to each from this classification of society into professions. The ESPRIT DE CORPS frequently overpowers the jealousy which exists between individuals, and pushes on to advantageous situations some of the more fortunate of the profession; whilst, on the other hand, any injury or insult offered to the weakest, is redressed or resented by the whole body. There are other advantages which are perhaps of more importance to the public. The numbers which compose the learned professions in England are so considerable, that a kind of public opinion is generated amongst them, which powerfully tends to repress conduct that is injurious either to the profession or to the public. Again, the mutual jealousy and rivalry excited amongst the whole body is so considerable, that although the rank and estimation which an individual holds in the profession may be most unfairly appreciated, by taking the opinion of his rival; yet few estimations will be found generally more correct than the opinion of a whole profession on the merits of any one of its body. This test is of great value to the public, and becomes the more so, in proportion to the difficulty of the study to which the profession is devoted. It is by availing themselves of it that men of sense and judgment, who have occasion for the services of professional persons, are, in a great measure, guided in their choice.

The pursuit of science does not, in England, constitute a distinct profession, as it does in many other countries. It is therefore, on that ground alone, deprived of many of the advantages which attach to professions. One of its greatest misfortunes arises from this circumstance; for the subjects on which it is conversant are so difficult, and require such unremitted devotion of time, that few who have not spent years in their study can judge of the relative knowledge of those who pursue them. It follows, therefore, that the public, and even that men of sound sense and discernment, can scarcely find means to distinguish between the possessors of knowledge, in the present day, merely elementary, and those whose acquirements are of the highest order. This remark applies with peculiar force to all the more difficult applications of mathematics; and the fact is calculated to check the energies of those who only look to reputation in England.

As there exists with us no peculiar class professedly devoted to science, it frequently happens that when a situation, requiring for the proper fulfilment of its duties considerable scientific attainments, is vacant, it becomes necessary to select from among amateurs, or rather from among persons whose chief attention has been bestowed on other subjects, and to whom science has been only an occasional pursuit. A certain quantity of scientific knowledge is of course possessed by individuals in many professions; and when added to the professional acquirements of the army, the navy, or to the knowledge of the merchant, is highly meritorious: but it is obvious that this may become, when separated from the profession, quite insignificant as the basis of a scientific reputation.

To those who have chosen the profession of medicine, a knowledge of chemistry, and of some branches of natural history, and, indeed, of several other departments of science, affords useful assistance. Some of the most valuable names which adorn the history of English science have been connected with this profession.

The causes which induce the selection of the clerical profession are not often connected with science; and it is, perhaps, a question of considerable doubt whether it is desirable to hold out to its members hopes of advancement from such acquirements. As a source of recreation, nothing can be more fit to occupy the attention of a divine; and our church may boast, in the present as in past times, that the domain of science has been extended by some of its brightest ornaments.

In England, the profession of the law is that which seems to hold out the strongest attraction to talent, from the circumstance, that in it ability, coupled with exertion, even though unaided by patronage, cannot fail of obtaining reward. It is frequently chosen as an introduction to public life. It also presents great advantages, from its being a qualification for many situations more or less remotely connected with it, as well as from the circumstance that several of the highest officers of the state must necessarily have sprung from its ranks.

A powerful attraction exists, therefore, to the promotion of a study and of duties of all others engrossing the time most completely, and which is less benefited than most others by any acquaintance with science. This is one amongst the causes why it so very rarely happens that men in public situations are at all conversant even with the commonest branches of scientific knowledge, and why scarcely an instance can be cited of such persons acquiring a reputation by any discoveries of their own.

But, however consistent other sciences may be with professional avocations, there is one which, from its extreme difficulty, and the overwhelming attention which it demands, can only be pursued with success by those whose leisure is undisturbed by other claims. To be well acquainted with the present state of mathematics, is no easy task; but to add to the powers which that science possesses, is likely to be the lot of but few English philosophers.

SECTION 2. OF NATIONAL ENCOURAGEMENT.

The little encouragement which at all previous periods has been afforded by the English Government to the authors of useful discoveries, or of new and valuable inventions, is justified on the following grounds:

1. The public, who consume the new commodity or profit by the new invention, are much better judges of its merit than the government can be.

2. The reward which arises from the sale of the commodity is usually much larger than that which government would be justified in bestowing; and it is exactly proportioned to the consumption, that is, to the want which the public feel for the new article.

It must be admitted that, as general principles, these are correct: there are, however, exceptions which flow necessarily from the very reasoning from which they were deduced. Without entering minutely into these exceptions, it will be sufficient to show that all abstract truth is entirely excluded from reward under this system. It is only the application of principles to common life which can be thus rewarded. A few instances may perhaps render this position more evident. The principle of the hydrostatic paradox was known as a speculative truth in the time of Stevinus; [About the year 1600] and its application to raising heavy weights has long been stated in elementary treatises on natural philosophy, as well as constantly exhibited in lectures. Yet, it may fairly be regarded as a mere abstract principle, until the late Mr. Bramah, by substituting a pump instead of the smaller column, converted it into a most valuable and powerful engine.—The principle of the convertibility of the centres of oscillation and suspension in the pendulum, discovered by Huygens more than a century and a half ago, remained, until within these few years, a sterile, though most elegant proposition; when, after being hinted at by Prony, and distinctly pointed out by Bonenberger, it was employed by Captain Kater as the foundation of a most convenient practical method of determining the length of the pendulum.—The interval which separated the discovery, by Dr. Black, of latent heat, from the beautiful and successful application of it to the steam engine, was comparatively short; but it required the efforts of two minds; and both were of the highest order.—The influence of electricity in producing decompositions, although of inestimable value as an instrument of discovery in chemical inquiries, can hardly be said to have been applied to the practical purposes of life, until the same powerful genius which detected the principle, applied it, by a singular felicity of reasoning, to arrest the corrosion of the copper-sheathing of vessels. That admirably connected chain of reasoning, the truth of which is confirmed by its very failure as a remedy, will probably at some future day supply, by its successful application, a new proof of the position we are endeavouring to establish.

[I am authorised in stating, that this was regarded by Laplace as the greatest of Sir Humphry Davy's discoveries. It did not fail in producing the effect foreseen by Sir H. Davy,—the preventing the corrosion of the copper; but it failed as a cure of the evil, by producing one of an OPPOSITE character; either by preserving too perfectly from decay the surface of the copper, or by rendering it negative, it allowed marine animals and vegetables to accumulate on its surface, and thus impede the progress of the vessel.]

Other instances might, if necessary, be adduced, to show that long intervals frequently elapse between the discovery of new principles in science and their practical application: nor ought this at all to surprise us. Those intellectual qualifications, which give birth to new principles or to new methods, are of quite a different order from those which are necessary for their practical application.

At the time of the discovery of the beautiful theorem of Huygens, it required in its author not merely a complete knowledge of the mathematical science of his age, but a genius to enlarge its boundaries by new creations of his own. Such talents are not always united with a quick perception of the details, and of the practical applications of the principles they have developed, nor is it for the interest of mankind that minds of this high order should lavish their powers on subjects unsuited to their grasp.

In mathematical science, more than in all

others, it happens that truths which are at one period the most abstract, and apparently the most remote from all useful application, become in the next age the bases of profound physical inquiries, and in the succeeding one, perhaps, by proper simplification and reduction to tables, furnish their ready and daily aid to the artist and the sailor.

It may also happen that at the time of the discovery of such principles, the mechanical arts may be too imperfect to render their application likely to be attended with success. Such was the case with the principle of the hydrostatic paradox; and it was not, I believe, until the expiration of Mr. Bramah's patent, that the press which bears his name received that mechanical perfection in its execution, which has deservedly brought it into such general use.

On the other hand, for one person who is blessed with the power of invention, many will always be found who have the capacity of applying principles; and much of the merit ascribed to these applications will always depend on the care and labour bestowed in the practical detail.

If, therefore, it is important to the country that abstract principles should be applied to practical use, it is clear that it is also important that encouragement should be held out to the few who are capable of adding to the number of those truths on which such applications are founded. Unless there exist peculiar institutions for the support of such inquirers, or unless the Government directly interfere, the contriver of a thaumatrope may derive profit from his ingenuity, whilst he who unravels the laws of light and vision, on which multitudes of phenomena depend, shall descend unrewarded to the tomb.

Perhaps it may be urged, that sufficient encouragement is already afforded to abstract science in our different universities, by the professorships established at them. It is not however in the power of such institutions to create; they may foster and aid the development of genius; and, when rightly applied, such stations ought to be its fair and honourable rewards. In many instances their emolument is small; and when otherwise, the lectures which are required from the professor are not perhaps in all cases the best mode of employing the energies of those who are capable of inventing.

I cannot resist the opportunity of supporting these opinions by the authority of one of the greatest philosophers of a past age, and of expressing my acknowledgments to the author of a most interesting piece of scientific biography. In the correspondence which terminated in the return of Galileo to a professorship in his native country, he remarks, "But, because my private lectures and domestic pupils are a great hinderance and interruption of my studies, I wish to live entirely exempt from the former, and in great measure from the latter."—LIFE OF GALILEO, p.18. And, in another letter to Kepler, he speaks with gratitude of Cosmo, the Grand Duke of Tuscany, who "has now invited me to attach myself to him with the annual salary of 1000 florins, and with the title of Philosopher and principal Mathematician to his Highness, without the duties of any office to perform, but with most complete leisure; so that I can complete my treatise on

Mechanics, &c."—p.31. [Life of Galileo, published by the Society for the Diffusion of Useful Knowledge.]

Surely, if knowledge is valuable, it can never be good policy in a country far wealthier than Tuscany, to allow a genius like Mr. Dalton's, to be employed in the drudgery of elementary instruction. [I utter these sentiments from no feelings of private friendship to that estimable philosopher, to whom it is my regret to be almost unknown, and whose modest and retiring merit, I may, perhaps, have the misfortune to offend by these remarks. But Mr. Dalton was of no party; had he ever moved in that vortex which has brought discredit, and almost ruin, on the Royal Society of England;—had he taken part with those who vote to each other medals, and, affecting to be tired of the fatigues of office, make to each other requisitions to retain places they would be most reluctant to quit; his great and splendid discovery would long since have been represented to government. Expectant mediocrity would have urged on his claims to remuneration, and those who covered their selfish purposes with the cloak of science, would have hastened to shelter themselves in the mantle of his glory.—But the philosopher may find consolation for the tardy approbation of that Society, in the applause of Europe. If he was insulted by their medal, he escaped the pain of seeing his name connected with their proceedings.] Where would have been the military renown of England, if, with an equally improvident waste of mental power, its institutions had forced the Duke of Wellington to employ his life in drilling recruits, instead of planning campaigns?

If we look at the fact, we shall find that the great inventions of the age are not, with us at least, always produced in universities. The doctrines of "definite proportions," and of the "chemical agency of electricity,"—principles of a high order, which have immortalized the names of their discoverers,—were not produced by the meditations of the cloister: nor is it in the least a reproach to those valuable institutions to mention truths like these. Fortunate circumstances must concur, even to the greatest, to render them eminently successful. It is not permitted to all to be born, like Archimedes, when a science was to be created; nor, like Newton, to find the system of the world "without form and void;" and, by disclosing gravitation, to shed throughout that system the same irresistible radiance as that with which the Almighty Creator had illumined its material substance. It can happen to but few philosophers, and but at distant intervals, to snatch a science, like Dalton, from the chaos of indefinite combination, and binding it in the chains of number, to exalt it to rank amongst the exact. Triumphs like these are necessarily "few and far between;" nor can it be expected that that portion of encouragement, which a country may think fit to bestow on science, should be adapted to meet such instances. Too extraordinary to be frequent, they must be left, if they are to be encouraged at all, to some direct interference of the government.

The dangers to be apprehended from such a specific interference, would arise from one, or several, of the following circumstances:—That class of society, from whom the government is selected, might not possess sufficient knowledge either to judge themselves, or know upon whose judgment to rely. Or the number of persons devoting themselves to science, might not be sufficiently large to have due weight in the expression of public opinion. Or, supposing this class to be large, it might not enjoy, in the estimation of the world, a sufficiently high character for independence. Should these causes concur in any country, it might become highly injurious to commit the encouragement of science to any department of the government. This reasoning does not appear to have escaped the penetration of those who advised the abolition of the late Board of Longitude.

The question whether it is good policy in the government of a country to encourage science, is one of which those who cultivate it are not perhaps the most unbiased judges. In England, those who have hitherto pursued science, have in general no very reasonable grounds of complaint; they knew, or should have known, that there was no demand for it, that it led to little honour, and to less profit.

That blame has been attributed to the government for not fostering the science of the country is certain; and, as far as regards past administrations, is, to a great extent, just; with respect to the present ministers, whose strength essentially depends on public opinion, it is not necessary that they should precede, and they cannot remain long insensible to any expression of the general feeling. But supposing science were thought of some importance by any administration, it would be difficult in the present state of things to do much in its favour; because, on the one hand, the higher classes in general have not a profound knowledge of science, and, on the other, those persons whom they have usually consulted, seem not to have given such advice as to deserve the confidence of government. It seems to be forgotten, that the money allotted by government to purposes of science ought to be expended with the same regard to prudence and economy as in the disposal of money in the affairs of private life.

[Who, for instance, could have advised the government to incur the expense of printing SEVEN HUNDRED AND FIFTY copies of the Astronomical Observations made at Paramatta, to form a third part of the Philosophical Transactions for 1829, whilst of the Observations made at the Royal Observatory at Greenwich, two hundred and fifty copies only are printed?

Of these seven hundred and fifty copies, seven hundred and ten will be distributed to members of the Royal Society, to six hundred of whom they will probably be wholly uninteresting or useless; and thus the country incurs a constantly recurring annual expense. Nor is it easy to see on what principle a similar destination could be refused for the observations made at the Cape of Good Hope.]

To those who measure the question of the national encouragement of science by its value in pounds, shillings, and pence, I will here state a fact, which, although pretty generally known, still, I think, deserves attention. A short time since it was discovered by government that the terms on which annuities had been granted by them were erroneous, and new tables were introduced by act of Parliament. It was stated at the time that the erroneous tables had caused a loss to the country of between two and three millions sterling. The fact of the sale of those annuities being a losing concern was long known to many; and the government appear to have been the last to be informed on the subject. Half the interest of half that loss, judiciously applied to the encouragement of mathematical science, would, in a few years, have rendered utterly impossible such expensive errors.

To those who bow to the authority of great names, one remark may have its weight. The MECANIQUE COELESTE, [The first volume of the first translation of this celebrated work into our own language, has just arrived in England from—America.] and the THEORIE ANALYTIQUE DES PROBABILITES, were both dedicated, by Laplace, to Napoleon. During the reign of that extraordinary man, the triumphs of France were as eminent in Science as they were splendid in arms. May the institutions which trained and rewarded her philosophers be permanent as the benefits they have conferred upon mankind!

In other countries it has been found, and is admitted, that a knowledge of science is a recommendation to public appointments, and that a man does not make a worse ambassador because he has directed an observatory, or has added by his discoveries to the extent of our knowledge of animated nature. Instances even are not wanting of ministers who have begun their career in the inquiries of pure analysis. As such examples are perhaps more frequent than is generally imagined, it may be useful to mention a few of those men of science who have formerly held, or who now hold, high official stations in the governments of their respective countries.

Country.	Name.	Department of Science.	Public Office.
France..	Marquis Laplace(1)	Mathematics	President of the Conservative Senate.
France..	M.Carnot	Mathematics	Minister of War.
France..	Count Chaptal(2)	Chemistry	Minister of the Interior.
France..	Baron Cuvier(3)	Comparative Anatomy, History	Minister of Public Instruction
Prussia..	Baron Humboldt	Oriental Languages	Ambassador to England
Prussia..	Baron Alexander	The celebrated	Chamberlain to

| Humboldt | Traveller | the King of Prussia |

Modena. Marquis Rangoni(4) Mathematics
Minister of

> Finance and
> of Public
> Instruction,
> President of
> Italian Academy
> of Forty.

Tuscany. Count Fossombroni   Mathematics
Prime Minister
(5)   of the Grand Duke
of Tuscany.

Saxony.. M. Lindenau(6)   Astronomy
Ambassador.

(1) Author of the MECANIQUE COELESTE.
(2) Author of TRAITE DE CHIMIE APPLIQUE
AUX ARTS.
(3) Author of LECONS D'ANATOMIE
COMPAREE—RECHERCHES SUR OSSEMENS
FOSSILES &c. &c.
(4) Author of MEMORIA SULLE FUNZIONI
GENERATRICI, Modena, 1824,
and of various other memoirs on mathematical
subjects.
(5) Author of several memoirs on mechanics and
hydraulics, in the
Transactions of the Academy of Forty.

(6) Author of TABLES BAROMETRIQUES, Gotha, 1809—TABULAE VENERIS, NOVAE ET CORRECTAE, Gothae, 1810—INVESTIGATIO NOVA ORBITAE A MERCURIO CIRCA SOLEM DESCRIPTAE, Gothae, 1813, and of other works.

M. Lindenau, the Minister from the King of Saxony to the King of the Netherlands, commenced his career as astronomer at the observatory of the Grand Duke of Gotha, by whom he was sent as his representative at the German Diet. On the death of the late reigning Duke, M. Lindenau was invited to Dresden, and filled the same situation under the King of Saxony; after which he was appointed his minister at the court of the King of the Netherlands. Such occurrences are not to be paralleled in our own country, at least not in modern times. Newton was, it is true, more than a century since, appointed Master of the Mint; but let any person suggest an appointment of a similar kind in the present day, and he will gather from the smiles of those to whom he proposes it that the highest knowledge conduces nothing to success, and that political power is almost the only recommendation.

SECTION 3. *Of Encouragement from Learned Societies.*

There are several circumstances which concur in inducing persons pursuing science, to unite together, to form societies or academies. In former times, when philosophical instruments were more rare, and the art of making experiments was less perfectly known, it was almost necessary. More recently, whilst numerous additions are constantly making to science, it has been found that those who are most capable of extending human knowledge, are frequently least able to encounter the expense of printing their investigations. It is therefore convenient, that some means should be devised for relieving them from this difficulty, and the volumes of the transactions of academies have accomplished the desired end.

There is, however, another purpose to which academies contribute. When they consist of a limited number of persons, eminent for their knowledge, it becomes an object of ambition to be admitted on their list. Thus a stimulus is applied to all those who cultivate science, which urges on their exertions, in order to acquire the wished-for distinction. It is clear that this envied position will be valued in proportion to the difficulty of its attainment, and also to the celebrity of those who enjoy it; and whenever the standard of scientific knowledge which qualifies for its ranks is lowered, the value of the distinction itself will be diminished. If, at any time, a multitude of persons having no sort of knowledge of science are admitted, it must cease to be sought after as an object of ambition by men of science, and the class of persons to whom it will become an object of desire will be less intellectual.

Let us now compare the numbers composing some of the various academies of Europe.-The Royal Society of London, the Institute of France, the Italian Academy of Forty, and the Royal Academy of Berlin, are amongst the most distinguished.

Name Country.	Population.	Number of Members of its Academy.	Number of Foreign Members
1. England.	22,299,000	685	50
2. France.	32,058,000	76	8 Mem. 100 Corr.
8. Prussia.	12,915,000	38	16
4. Italy..	12,000,000	40	8

It appears then, that in France, one person out of 427,000 is a member of the Institute. That in Italy and Prussia, about one out of 300,000 persons is a member of their Academies. That in England, every 32,000 inhabitants produces a Fellow of the Royal Society. Looking merely at these proportions, the estimation of a seat in the Academy of Berlin, must be more than nine times as valuable as a similar situation in England; and a member of the Institute of France will be more than thirteen times more rare in his country than a Fellow of the Royal Society is in England.

Favourable as this view is to the dignity of such situations in other countries, their comparative rarity is by no means the most striking difference in the circumstances of men of science. If we look at the station in society occupied by the SAVANS of other countries, in several of them we shall find it high, and their situations profitable. Perhaps, at the present moment, Prussia is, of all the countries in Europe, that which bestows the greatest attention, and most unwearied encouragement on science. Great as are the merits of many of its philosophers, much of this support arises from the character of the reigning family, by whose enlightened policy even the most abstract sciences are fostered.

The maxim that "knowledge is power," can be perfectly comprehended by those only who are themselves well versed in science; and to the circumstance of the younger branches of the royal family of Prussia having acquired considerable knowledge in such subjects, we may attribute the great force with which that maxim is appreciated.

In France, the situation of its SAVANS is highly respectable, as well as profitable. If we analyze the list of the Institute, we shall find few who do not possess titles or decorations; but as the value of such marks of royal favour must depend, in a great measure, on their frequency, I shall mention several particulars which are probably not familiar to the English reader. [This analysis was made by comparing the list of the Institute, printed for that body in 1827, with the ALMANACH ROYALE for

Number of the Members of the   Total Number
of each Class
Institute of France who belong   of the Legion of
Honour.
to the Legion of Honour.

GrandCroix.........	3	80
GrandOfficier.....	3	160
Commandeur........	4	400
Officier..........	17	2,000
Chevalier.........	40	Not limited.

Number of Members of the Institute   Total Number
decorated with   of
the Order of St. Michel.   that Order.

Grand Croix.......	2	
		100
Chevalier.........	27	

Amongst the members of the Institute there
are,—

Dukes...................	2
Marquis................	1
Counts.................	4
Viscounts..............	2
Barons.................	14
	—23

Of these there are
Peers of France.......... 5

We might, on turning over the list of the 685 members of the Royal Society, find a greater number of peers than there are in the Institute of France; but a fairer mode of instituting the comparison, is to inquire how many titled members there are amongst those who have contributed to its Transactions. In 1827, there were one hundred and nine members who had contributed to the Transactions of the Royal Society; amongst these were found:—

Peer........................	1
Baronets....................	5
Knights.....................	5

It should be observed, that five of these titles were the rewards of members of the medical profession, and one only, that of Sir H. Davy, could be attributed exclusively to science.

It must not be inferred that the titles of nobility in the French list, were all of them the rewards of scientific eminence; many are known to have been such; but it would be quite sufficient for the argument to mention the names of Lagrange, Laplace, Berthollet, and Chaptal.

The estimation in which the public hold literary claims in France and England, was curiously illustrated by an incidental expression in the translation of the debates in the House of Lords, on the occasion of His Majesty's speech at the commencement of the session of 1830. The Gazette de France stated, that the address was moved by the Duc de Buccleugh, "CHEF DE LA MAISON DE WALTER SCOTT." Had an English editor wished to particularize that nobleman, he would undoubtedly have employed the term WEALTHY, or some other of the epithets characteristic of that quality most esteemed amongst his countrymen.

If we turn, on the other hand, to the emoluments of science in France, we shall find them far exceed those in our own country. I regret much that I have mislaid a most interesting memorandum on this subject, which I made several years since: but I believe my memory on the point will not be found widely incorrect. A foreign gentleman, himself possessing no inconsiderable acquaintance with science, called on me a few years since, to present a letter of introduction. He had been but a short time in London; and, in the course of our conversation, it appeared to me that he had imbibed very inaccurate ideas respecting our encouragement of science.

Thinking this a good opportunity of instituting a fair comparison between the emoluments of science in the two countries, I placed a sheet of paper before him, and requested him to write down the names of six Englishmen, in his opinion, best known in France for their scientific reputation. Taking another sheet of paper, I wrote upon it the names of six Frenchmen, best known in England for their scientific discoveries. We exchanged these lists, and I then requested him to place against each name (as far as he knew) the annual income of the different appointments held by that person. In the mean time, I performed the same operation on his list, against some names of which I was obliged to place a ZERO. The result of the comparison was an average of nearly 1200L. per annum for the six French SAVANS whom I had named. Of the average amount of the sums received by the English, I only remember that it was very much smaller. When we consider what a command over the necessaries and luxuries of life 1200L. will give in France, it is underrating it to say it is equal to 2000L. in this country.

Let us now look at the prospects of a young man at his entrance into life, who, impelled by an almost irresistible desire to devote himself to the abstruser sciences, or who, confident in the energy of youthful power, feels that the career of science is that in which his mental faculties are most fitted to achieve the reputation for which he pants. What are his prospects? Can even the glowing pencil of enthusiasm add colour to the blank before him? There are no situations in the state; there is no position in society to which hope can point, to cheer him in his laborious path. If, indeed, he belong to one of our universities, there are some few chairs in his OWN Alma Mater to which he may at some distant day pretend; but these are not numerous; and whilst the salaries attached are seldom sufficient for the sole support of the individual, they are very rarely enough for that of a family. What then can he reply to the entreaties of his friends, to betake himself to some business in which perhaps they have power to assist him, or to choose some profession in which his talents may produce for him their fair reward? If he have no fortune, the choice is taken away: he MUST give up that line of life in which his habits of thought and his ambition qualify him to succeed eminently, and he MUST choose the bar, or some other profession, in which, amongst so many competitors, in spite of his great talents, he can be but moderately successful. The loss to him is great, but to the country it is greater. We thus, by a destructive misapplication of talent which our institutions create, exchange a profound philosopher for but a tolerable lawyer.

If, on the other hand, he possess some moderate fortune of his own; and, intent on the glory of an immortal name, yet not blindly ignorant of the state of science in this country, he resolve to make for that aspiration a sacrifice the greater, because he is fully aware of its extent;—if, so circumstanced, he give up a business or a profession on which he might have entered with advantage, with the hope that, when he shall have won a station high in the ranks of European science, he may a little augment his resources by some of those few employments to which science leads;—if he hope to obtain some situation, (at the Board of Longitude, for example,) [This body is now dissolved] where he may be permitted to exercise the talents of a philosopher for the paltry remuneration of a clerk, he will find that other qualifications than knowledge and a love of science are necessary for its attainment. He will also find that the high and independent spirit, which usually dwells in the breast of those who are deeply versed in these pursuits, is ill adapted for such appointments; and that even if successful, he must hear many things he disapproves, and raise no voice AGAINST them.

Thus, then, it appears that scarcely any man can be expected to pursue abstract science unless he possess a private fortune, and unless he can resolve to give up all intention of improving it. Yet, how few thus situated are likely to undergo the labour of the acquisition; and if they do from some irresistible impulse, what inducement is there for them to deviate one step from those inquiries in which they find the greatest delight, into those which might be more immediately useful to the public?

CHAPTER III. GENERAL STATE OF LEARNED SOCIETIES IN ENGLAND.

The progress of knowledge convinced the world that the system of the division of labour and of cooperation was as applicable to science, as it had been found available for the improvement of manufactures. The want of competition in science produced effects similar to those which the same cause gives birth to in the arts. The cultivators of botany were the first to feel that the range of knowledge embraced by the Royal Society was too comprehensive to admit of sufficient attention to their favourite subject, and they established the Linnean Society. After many years, a new science arose, and the Geological Society was produced. At an another and more recent epoch, the friends of astronomy, urged by the wants of their science, united to establish the Astronomical Society. Each of these bodies found, that the attention devoted to their science by the parent establishment was insufficient for their wants, and each in succession experienced from the Royal Society the most determined opposition.

Instituted by the most enlightened philosophers, solely for the promotion of the natural sciences, that learned body justly conceived that nothing could be more likely to render these young institutions permanently successful, than discouragement and opposition at their commencement. Finding their first attempts so eminently successful, they redoubled the severity of their persecution, and the result was commensurate with their exertions, and surpassed even their wildest anticipations. The Astronomical Society became in six years known and respected throughout Europe, not from the halo of reputation which the glory of its vigourous youth had thrown around the weakness of its declining years; but from the sterling merit of "its unpretending deeds, from the sympathy it claimed and received from every practical astronomer, whose labours it relieved, and whose calculations it lightened."

But the system which worked so well is now changed, and the Zoological and Medico-Botanical Societies were established without opposition: perhaps, indeed, the total failure of the latter society is the best proof of the wisdom which guided the councils of the Royal. At present, the various societies exist with no feelings of rivalry or hostility, each pursuing its separate objects, and all uniting in deploring with filial regret, the second childhood of their common parent, and the evil councils by which that sad event has been anticipated.

It is the custom to attach certain letters to the names of those who belong to different societies, and these marks of ownership are by many considered the only valuable part of their purchase on entry. The following is a list of some of these societies. The second column gives the ready-money prices of the tail-pieces indicated in the third.

SOCIETIES.	Fees on Admission including Composition for Annual Payments.	Letters Appended
	L. s. d.	
Royal Society.............	50 0 0	F.R.S.
Royal Society of Edinburgh.	25 4 0*	F.R.S.E.
Royal Academy of Dublin...	26 5 0	M.R.I.A.
Royal Society of Literature	36 15 0	F.R.S.Lit.
Antiquarian................	50 8 0	F.A.S.
Linnean...................	36 0 0	F.L.S.
Geological................	34 15 0	F.G.S.
Astronomical..............	25 4 0	M.A.S.
Zoological................	26 5 0	F.Z.S.
Royal Institution.........	50 0 0	M.R.I.
Royal Asiatic.............	31 10 0	F.R.A.S.
Horticultural.............	43 6 0	F.H.S.
Medico-Botanical..........	21 0 0	F.M.B.S.

[* The Royal Society of Edinburgh now requires, for composition in lieu of annual contributions, a sum dependent on the value of the life of the member.]

Thus, those who are ambitious of scientific distinction, may, according to their fancy, render their name a kind of comet, carrying with it a tail of upwards of forty letters, at the average cost of 10L. 9s. 9d. per letter.

Perhaps the reader will remark, that science cannot be declining in a country which supports so many institutions for its cultivation. It is indeed creditable to us, that the greater part of these societies are maintained by the voluntary contributions of their members. But, unless the inquiries which have recently taken place in some of them should rectify the SYSTEM OF MANAGEMENT by which several have been oppressed, it is not difficult to predict that their duration will be short. Full PUBLICITY, PRINTED STATEMENTS OF ACCOUNTS, and occasional DISCUSSIONS and inquiries at GENERAL MEETINGS, are the only safeguards; and a due degree of VIGILANCE should be exercised on those who DISCOURAGE these principles. Of the Royal Society, I shall speak in a succeeding page; and I regret to add, that I might have said more. My object is to amend it; but, like all deeply-rooted complaints, the operation which alone can contribute to its cure, is necessarily painful. Had the words of remonstrance or reproof found utterance through other channels, I had gladly been silent, content to support by my vote the reasonings of the friend of science and of the Society. But this has not been the case, and after frustrated efforts to introduce improvements, I shall now endeavour, by the force of plain, but perhaps painful truths, to direct public opinion in calling for such a reform, as shall rescue the Royal Society from contempt in our own country, from ridicule in others.

On the next five societies in the list, I shall offer no remarks. Of the Geological, I shall say a few words. It possesses all the freshness, the vigour, and the ardour of youth in the pursuit of a youthful science, and has succeeded in a most difficult experiment, that of having an oral discussion on the subject of each paper read at its meetings. To say of these discussions, that they are very entertaining, is the least part of the praise which is due to them. They are generally very instructive, and sometimes bring together isolated facts in the science which, though insignificant when separate, mutually illustrate each other, and ultimately lead to important conclusions. The continuance of these discussions evidently depends on the taste, the temper, and the good sense of the speakers. The things to be avoided are chiefly verbal criticisms—praise of each other beyond its reasonable limits, and contest for victory. This latter is, perhaps, the most important of the three, both for the interests of the Society and of truth. With regard to the published volumes of their Transactions, it may be remarked, that if members were in the habit of communicating their papers to the Society in a more finished state, it would be attended with several advantages; amongst others, with that of lightening the heavy duties of the officers, which are perhaps more laborious in this Society than in most others. To court publicity in their accounts and proceedings, and to endeavour to represent all the feelings of the Society in the Council, and to avoid permanent Presidents, is a recommendation not peculiarly addressed to this Society, but would contribute to the well-being of all.

Of the Astronomical Society, which, from the nature of its pursuits, could scarcely admit of the discussions similar to those of the Geological, I shall merely observe, that I know of no secret which has caused its great success, unless it be attention to the maxims which have just been stated.

On the Zoological Society, which affords much rational amusement to the public, a few hints may at present suffice. The largeness of its income is a frightful consideration. It is too tempting as the subject for jobs, and it is too fluctuating and uncertain in its amount, not to render embarrassment in the affairs of the Society a circumstance likely to occur, without the greatest circumspection. It is most probable, from the very recent formation of this Institution, that its Officers and Council are at present all that its best friends could wish; but it is still right to mention, that in such a Society, it is essentially necessary to have men of business on the Council, as well as persons possessing extensive knowledge of its pursuits. It is more dangerous in such a Society than in any other, to pay compliments, by placing gentlemen on the Council who have not the qualifications which are requisite; a frequent change in the members of the Council is desirable, in order to find out who are the most regular attendants, and most qualified to conduct its business. Publicity in its accounts and proceedings is, from the magnitude of its funds, more essential to the Zoological than to any other society; and it is rather a fearful omen, that a check was attempted to be given to such inquiries at the last anniversary meeting. If it is to be a scientific body, the friends of science should not for an instant tolerate such attempts.

It frequently happens, that gentlemen take an active part in more than one scientific society: in that case, it may be useful to derive instruction as to their merits, by observing the success of their measures in other societies.

The Asiatic Society has, amongst other benefits, caused many valuable works to be translated, which could not have otherwise been published.

The Horticultural Society has been ridden almost to death, and is now rousing itself; but its constitution seems to have been somewhat impaired. There are hopes of its purgation, and ultimate restoration, notwithstanding a debt of 19,000L., which the Committee of Inquiry have ascertained to exist. This, after all, will not be without its advantage to science, if it puts a stop to HOUSE-LISTS, NAMED BY ONE OR TWO PERSONS,—to making COMPLIMENTARY councillors,—and to auditing the accounts WITHOUT EXAMINING EVERY ITEM, or to omitting even that form altogether.

The Medico-Botanical Society suddenly claimed the attention of the public; its pretensions were great—its assurance unbounded. It speedily became distinguished, not by its publications or discoveries, but by the number of princes it enrolled in its list. It is needless now to expose the extent of its short-lived quackery; but the evil deeds of that institution will long remain in the impression they have contributed to confirm throughout Europe, of the character of our scientific establishments. It would be at once a judicious and a dignified course, if those lovers of science, who have been so grievously deceived in this Society, were to enrol upon the latest page of its history its highest claim to public approbation, and by signing its dissolution, offer the only atonement in their power to the insulted science of their country. As with a singular inversion of principle, the society contrived to render EXPULSION* the highest HONOUR it could confer; so it remains for it to exemplify, in suicide, the sublimest virtue of which it is capable. [* They expelled from amongst them a gentleman, of whom it is but slight praise to say, that he is the first and most philosophical botanist of our own country, and who is admired abroad as he is respected at home. The circumstance which surprised the world was not his exit from, but his previous entrance into that Society.]

CHAPTER IV. STATE OF THE ROYAL SOCIETY IN PARTICULAR.

As the venerable first parent of English, and I might perhaps say, of European scientific societies; as a body in the welfare of which, in the opinions of many, the interests of British science are materially involved, I may be permitted to feel anxiously, and to speak more in detail.

SECTION 1. MODE OF BECOMING A FELLOW OF THE ROYAL SOCIETY.

I have no intention of stating what ought to be the qualifications of a Fellow of the Royal Society; but, for years, the practical mode of arriving at that honour, has been as follows:—

A. B. gets any three Fellows to sign a certificate, stating that he (A. B.) is desirous of becoming a member, and likely to be a useful and valuable one. This is handed in to the Secretary, and suspended in the meeting-room. At the end of ten weeks, if A. B. has the good fortune to be perfectly unknown by any literary or scientific achievement, however small, he is quite sure of being elected as a matter of course. If, on the other hand, he has unfortunately written on any subject connected with science, or is supposed to be acquainted with any branch of it, the members begin to inquire what he has done to deserve the honour; and, unless he has powerful friends, he has a fair chance of being black-balled. [I understand that certificates are now read at the Council, previously to their being hung up in the meeting-room; but I am not aware that this has in the slightest degree diminished their number, which was, at the time of writing this note, TWENTY-FOUR.]

In fourteen years' experience, the few whom I have seen rejected, have all been known persons; but even in such cases a hope remains;—perseverance will do much, and a gentleman who values so highly the distinction of admission to the Royal Society, may try again; and even after being twice black-balled, if he will a third time condescend to express his desire to become a member, he may perhaps succeed, by the aid of a hard canvass. In such circumstances, the odds are much in favour of the candidate possessing great scientific claims; and the only objection that could then reasonably be suggested, would arise from his estimating rather too highly a distinction which had become insignificant from its unlimited extension.

It should be observed, that all members contribute equally, and that the sum now required is fifty pounds. It used, until lately, to be ten pounds on entrance, and four pounds annually. The amount of this subscription is so large, that it is calculated to prevent many men of real science from entering the Society, and is a very severe tax on those who do so; for very few indeed of the cultivators of science rank amongst the wealthy classes. Several times, whilst I have been consulting books or papers at Somerset House, persons have called to ask the Assistant-secretary the mode of becoming a member of the Royal Society. I should conjecture, from some of these applications, that it is not very unusual for gentlemen in the country to order their agents in London to take measures for putting them up at the Royal Society.

SECTION 2. OF THE PRESIDENCY AND VICE-PRESIDENCIES.

Why Mr. Davies Gilbert became President of the Royal Society I cannot precisely say. Let him who penned, and those who supported this resolution solve the enigma:
"It was Resolved,
"That it is the opinion of the Council that Davies Gilbert, Esq. is by far the most fit person to be proposed to the Society at the approaching anniversary as President, and that he be

recommended accordingly."

To resolve that he was a FIT person might have been sufficiently flattering: to state that he was the most fit, was a little hard upon the rest of the Society; but to resolve that he was "BY FAR THE MOST FIT" was only consistent with that strain of compliment in which his supporters indulge, and was a eulogy, by no means unique in its kind, I believe, even at that very Council.

That Mr. Gilbert is a most amiable and kind-hearted man will be instantly admitted by all who are, in the least degree, acquainted with him: that he is fit for the chair of the Royal Society, will be allowed by few, except those who have committed themselves to the above-quoted resolution.

Possessed of knowledge and of fortune more than sufficient for it, he might have been the restorer of its lustre. He might have called round him, at the council board, those most actively engaged in the pursuits of science, most anxious for the improvement of the Royal Society. Instead of himself proposing resolutions, he might have been, what a chairman ought to be, the organ of the body over which he presides. By the firmness of his own conduct he might have taught the subordinate officers of the Society the duties of their station. Instead of paying compliments to Ministers, who must have smiled at his simplicity, he might have maintained the dignity of his Council by the dignity of knowledge.

But he has chosen a different path; with no motives of interest to allure, or of ambition to betray him, instead of making himself respected as the powerful chief of a united republic,—that of science, —he has grasped at despotic power, and stands the feeble occupant of its desolated kingdom, trembling at the force of opinions he might have directed, and refused even the patronage of their names by those whose energies he might have commanded.

Mr. Gilbert told the Society he accepted the situation for a year; and this circumstance caused a difficulty in finding a Treasurer: an office which he had long held, and to which he wished to return.

Another difficulty might have arisen, from the fact of the late Board of Longitude comprising amongst its Members the PRESIDENT of the Royal Society, and three of its Fellows, appointed by the President and Council. Of course, when Mr. Gilbert accepted the higher situation, he became, EX OFFICIO, a Member of the Board of Longitude; and a vacancy occurred, which ought to have been filled up by the President and Council. But when this subject was brought before them, in defiance of common sense, and the plain meaning of the act of parliament, which had enacted that the Board of Longitude should have the assistance of four persons belonging to the Royal Society, Mr. Gilbert refused to allow it to be filled up, on the ground that he should not be President next year, and had made no vacancy.

Next year Mr. Gilbert wished again to be President one other year; but the Board of Longitude was dissolved, otherwise we might have had some LOCUM TENENS to retire at Mr. Gilbert's pleasure.

These circumstances are in themselves of trifling importance, but they illustrate the character of the proceedings: and it is not becoming the dignity of science or of the Society that its officers should be so circumstanced as to have an apparent and direct interest in supporting the existing President, in order to retain their own places; and if such a system is once discovered, doubt immediately arises as to the frequency of such arrangements.

SECTION 3. OF THE SECRETARISHIPS.

Whether the present Secretaries are the best qualified to aid in reforming the Society, is a question I shall not discuss. With regard to the senior Secretary, the time of his holding office is perhaps more unfortunate than the circumstance. If I might be permitted to allude for a moment to his personal character, I should say that the mild excellencies of his heart have prevented the Royal Society from deriving the whole of that advantage from his varied knowledge and liberal sentiments which some might perhaps have anticipated; and many will agree with me in regretting that his judgment has not directed a larger portion of the past deeds of the Councils of the Royal Society. Of the junior Secretary I shall only observe, that whilst I admit his industry, his perseverance, and his talents, I regret to see such valuable qualities exerted at a disadvantage, and that

I sincerely wish them all the success they merit in situations more adapted for their developement.

There are, however, some general principles which it may be important to investigate, which relate to the future as well as to the past state of the office of Secretary of the Royal Society. Inconvenience has already arisen from having had at a former period one of our Secretaries the conductor of a scientific journal; and this is one of the points in which I can agree with those who now manage the affairs of the Society. [These observations were written previous to the late appointment, to which I now devote Section 6. Experience seems to be lost on the Council of the Royal Society.] Perhaps it might be advantageous to extend the same understanding to the other officers of the Society at least, if not to the members of its Council.

Another circumstance worthy of the attention of the Society is, to consider whether it is desirable, except in special cases, to have military persons appointed to any of its offices. There are several peculiarities in the military character, which, though they do not absolutely unfit their possessors for the individual prosecution of science, may in some degree disqualify such persons from holding offices in scientific institutions. The habits both of obedience and command, which are essential in military life, are little fitted for that perfect freedom which should reign in the councils of science. If a military chief commit an oversight or an error, it is necessary, in order to retain the confidence of those he commands, to conceal or mask it as much as possible. If an experimentalist make a mistake, his only course to win the confidence of his fellow-labourers in science, and to render his future observations of any use, is to acknowledge it in the most full and explicit manner. The very qualifications which contribute to the professional excellence of the soldier, constitute his defects when he enters the paths of science; and it is only in those rare cases where the force of genius is able to control and surmount these habits, that his admission to the offices of science can be attended with any advantage to it.

Another objection deserving notice, although not applying exclusively to the military profession, is, that persons not imbued with the feelings of men of science, when they have published their observations, are too apt to view every criticism upon them as a personal question, and to consider that it is as offensive to doubt the accuracy of their observations as it is to doubt their word. Nothing can be more injurious to science than that such an opinion should be tolerated. The most unreserved criticism is necessary for truth; and those suspicions respecting his own accuracy, which every philosophical experimenter will entertain concerning his own researches, ought never to be considered as a reproach, when they are kept in view in examining the experiments of others. The minute circumstances and apparently trivial causes which lend their influence towards error, even in persons of the most candid judgment, are amongst the most curious phenomena of the human mind.

The importance of affording every aid to enable others to try the merits of observations, has been so well expressed by Mayer, that I shall conclude these remarks with an extract from the Preface to his Observations:

"Officii enim cujusque observatoris ease reor, de habitu instrumenti sui, de cura ac precautione, qua usus est, ad illud recte tractandum, deque mediis in errores ejus inquirendi rationem reddere publice, ut aliis quoque copia sit judicandi, quanta fides habenda conclusionibus ex nostris observationibus deductis aut deducendis. Hoc cum minus fecissent precedentis saeculi astronomi, praxin nimis secure, nimisque theoretice tractantes, factum inde potissimum est, ut illorum observationes tot vigiliis tantoque labore comparatae tam cito obsoleverint." P. viii.

There are certain duties which the Royal Society owes to its own character as well as to the public, which, having been on some occasions apparently neglected, it may be here the proper place to mention, since it is reasonable to suppose that attention to them is within the province of its Secretaries.

The first to which I shall allude is the singular circumstances attending the fact of the Royal Society having printed a volume of Astronomical Observations which were made at the Observatory of Paramatta (New South Wales), bearing the title of "The Third Part of the Philosophical Transactions for the Year 1829."

Now this Observatory was founded at the private expense of a British officer; the instruments were paid for out of his purse; two observers were brought from Europe, to be employed in making use of those instruments, at salaries defrayed by him. A considerable portion of the observations so printed were made by these astronomers during their employment in his service, and some of them are personally his own. Yet has the Royal Society, in adopting them as part of its Transactions, omitted all mention, either in their title-page, preface, or in any part of the volume, of the FACT that the world owed these valuable observations to the enlightened munificence of Lieutenant-General Sir Thomas Brisbane; whose ardent zeal in the pursuit of science induced him to found, at his own private expense, an establishment which it has been creditable to the British Government to continue as a national institution. Had any kindred feelings existed in the Council, instead of endeavouring to shift the responsibility, they would have hastened to rectify an omission, less unjust to the individual than it was injurious to English science.

Another topic, which concerns most vitally the character and integrity of the Royal Society, I hardly know how to approach. It has been publicly stated that confidence cannot be placed in the written minutes of the Society; and an instance has been adduced, in which an entry has been asserted to have been made, which could not have been the true statement of what actually passed at the Council.

The facts on which the specific instance rests are not difficult to verify by members of the Royal Society. I have examined them, and shall state them before I enter on the reasoning which may be founded upon them. In the minutes of the Council, 26th November, 1829, we find—

"Resolved, that the following gentlemen be recommended to be put upon the Council for the ensuing year." [Here follows a list of persons, amongst whom the name of Sir John Franklin occurs [Sir John Franklin was absent from London, and altogether unacquainted with this transaction, until he saw it stated in the newspapers some months after it had taken place. That his name was the one substituted for that of Captain Beaufort I know, from other evidence which need not be produced here, as the omission of the latter name is the charge that has been made.], and that of Captain Beaufort is not found. [Any gentleman may satisfy himself that this is not a mistake of the Assistant Secretary's, in copying, by consulting the rough minutes of that meeting of the Council, which it might perhaps be as well to write in a rough minute-book, instead of upon loose sheets of paper; nor can it be attributed to any error arising from accidentally mislaying the real minutes, for in that case the error would have been rectified immediately it was detected; and this has remained uncorrected, although publicly spoken of for months. As there is no erasure in the list, one is reluctantly compelled to conjecture that the real minutes of that meeting have been destroyed.]]

Now this could not be the list actually recommended by the Council on the morning of the 26th of November, because the President himself, on the evening of that day, informed Capt. Beaufort that he was placed on the house list; and that officer, with the characteristic openness of his profession, wrote on the next or the following day to the President, declining that situation, and stating his reasons for the step.

Upon the fact, therefore, of the suppression of part of a resolution of the Council, on the 26th of November, there can be no doubt; but in order to understand the whole nature of the transaction, other information is necessary. It has been the wish of many members of the Society, that the President should not absolutely name his own Council, but that the subject should be discussed fairly at the meeting previous to the Anniversary—this has always been opposed by Mr. Gilbert, and those who support him. Now, it has been stated, that, at the meeting of the Council on the 26th of November, the President took out of his pocket a bit of paper, from which he read the names of several persons as fit to be on the Council for the ensuing year;—that it was not understood that any motion was made, and it is certain that none was seconded, nor was any ballot taken on such an important question; and it was a matter of considerable surprise to some of those present, to discover afterwards that it was entered on the minutes as a resolution. This statement I have endeavoured to verify, and I believe it to be substantially correct; if it was a resolution, it was dictated, not discussed. It is also important to observe, that no similar resolution stands on the council-books for any previous year.

On examining the minutes of the succeeding Council, no notice of the letter of Captain Beaufort to the President is found. Why was it omitted? If the first entry had been truly made, there would have been no necessity for the omission; and after the insertion of that letter, a resolution would naturally have followed, recommending another name instead of the one withdrawn. Such was the natural and open course; but this would have exposed to the Society the weakness of those who manage it. If the rough minutes of each meeting of the Council were read over before it separated, and were copied previously to the next meeting, such a substitution could hardly have occurred; but, unfortunately, this is not the case, and the delay is in some cases considerable. Thus, the minutes of the three Councils, held on February 4, on February 11, and on March 11, were not entered on the minute-books of the Council on Tuesday, the 16th March; nor was this the fault of the Assistant-secretary, for up to that day the rough minutes of no one of those Councils had been transmitted to him.

Deeply as every friend to the Royal Society must regret such an occurrence, one slight advantage may accrue. Should that resolution be ever quoted hereafter to prove that the Council of 1829 really discussed the persons to be recommended as their successors, the detection of this suppression of one portion of it, will furnish better means of estimating the confidence due to the whole.

SECTION 4. OF THE SCIENTIFIC ADVISERS.

Whether it was feared by the PARTY who govern the Royal Society, that its Council would not be sufficiently tractable, or whether the Admiralty determined to render that body completely subservient to them, or whether both these motives concurred, I know not; but, low as has been for years its character for independence, and fallen as the Royal Society is in public estimation, it could scarcely be prepared for this last insult. In order to inform the public and the Society, (for I believe the fact is known to few of the members,) it will be necessary to trace the history of those circumstances which led to the institution of the offices of Scientific Advisers, from the time of the existence of the late Board of Longitude.

That body consisted, according to the act of parliament which established it, of certain official members, who usually possessed no knowledge of the subjects it was the duty of the Board to discuss— of certain professors of the two universities, and the Astronomer Royal, who had some knowledge, and who were paid 100L. a year for their attendance;—of three honorary members of the Royal Society, who combined the qualifications of the two preceding classes; and, lastly, of "three other persons," named

Resident Commissioners, who were supposed to be "WELL VERSED IN THE SCIENCES OF MATHEMATICS, ASTRONOMY, OR NAVIGATION," and who were paid a hundred a year to do the work of the Board.

The first three classes were permanent members, but the "three other persons" only held the appointment for ONE YEAR, and were renewable at the pleasure of the Admiralty. This Board was abolished by another act of parliament, on the ground that it was useless. Shortly after, the Secretary of the Admiralty communicated to the Council of the Royal Society, the copy of an Order in Council:

ADMIRALTY OFFICE,
November 1, 1828.

SIR, I am commanded by my Lords Commissioners of the Admiralty, to send herewith, for the information of the President and Council of the Royal Society, a copy of His Majesty's Order in Council of the 27th of last month; explaining that the salaries heretofore allowed to the Resident Commissioners of the Board of Longitude, and to the Superintendents of the Nautical Almanac, and of Chronometers, shall be continued to them, notwithstanding the abolition of the Board of Longitude. And I am to acquaint you, that the necessary orders have been given to the Navy Board for the payment of the said salaries.

I am, Sir,
Your most obedient humble servant,
JOHN BARROW. AT THE COURT AT WINDSOR,

27th October, 1828.

PRESENT,

The King's most Excellent Majesty in Council,

Whereas, there was this day read at the Board a Memorial from the Right Honourable the Lords Commissioners of the Admiralty, dated 4th of this instant, in the words following, viz.—

Whereas, by an Act of the 58th of his late Majesty's reign, cap. 20, instituted "An Act for the more effectually discovering the Longitude at sea, and encouraging attempts to find a Northern passage between the Atlantic and Pacific Oceans, and to approach the North Pole," three persons well versed in the sciences of Mathematics, Astronomy, or Navigation, were appointed as a Resident Committee of the Board of Commissioners for discovery of the Longitude at sea, and a Superintendent of the Nautical Almanac and of Chronometers was also appointed, with such salaries for the execution of those services as his Majesty might, by any Order in Council, be pleased to direct; and, whereas, your Majesty was in consequence, by your Order in Council of the 27th of May, 1828, most graciously pleased to direct, that the three said Resident Commissioners should be paid at the rate of 100L. a year each; and by your further Order in Council, of the 31st October, 1818, that the Superintendent of the Nautical Almanac should be allowed a salary of 300L., and the Superintendent of Chronometers 100L. a year; and, whereas, the act above mentioned has been repealed, and the Board of Longitude abolished; and doubts have therefore arisen, whether the said Orders in Council shall still continue in force; and whereas it is expedient that the said appointments be continued; We beg leave most humbly to submit to your Majesty, that your Majesty may be graciously pleased, by your Order in Council, to direct that the said offices of Superintendent of the Nautical Almanac, and of Superintendent of Chronometers; and also the three persons before-mentioned as a Resident Committee, to advise with the Commissioners for executing the Office of Lord High Admiral, on all questions of discoveries, inventions, calculations, and other scientific subjects, be continued, with the same duties and salaries, and under the same regulations as heretofore; and further beg most humbly to propose, that such three persons to form the Resident Committee, be chosen annually by the Commissioners for executing the office of Lord High Admiral, from among the Council of the Royal Society.

His Majesty, having taken the said Memorial into consideration, was pleased, by and with the advice of his Privy Council, to approve thereof and the Right Honourable the Lords Commissioners of the Admiralty are to give the necessary directions herein accordingly.

(Signed) JAMES HILLER.

Thus, it appeared that the Admiralty were to choose three persons from among the Council of the Royal Society, who were to have a hundred a year each during the pleasure of the Admiralty.

Such an open attack on the independence of the Council could not escape the remarks of some of the members, and a kind of mild remonstrance was made, in which the real ground of complaint was omitted.

MINUTE OF COUNCIL OF THE ROYAL SOCIETY. December 18, 1823.

RESOLVED, That in acknowledging the communication of the Lords Commissioners of the Admiralty, made to the Council of the Royal Society, on the 20th of November last, it be represented to them that inconvenience may arise from the plan therein specified, from the circumstance of all the members of the Council being annually elected by the Society at large; and that body being consequently subject to continual changes from year to year.

This was answered by the following letter from the Secretary of the Admiralty:

ADMIRALTY OFFICE, DEC. 30, 1828.

SIR, Having submitted to my Lords Commissioners of the Admiralty your Letter of the 18th instant, subjoining an extract from the Minutes of the proceedings of the Council of the Royal Society, arising out of the communication made to them by their Lordships, on the subject of his Majesty's Order in Council, of the fifth of October last, I have their Lordships' command to acquaint you, for the information of the President and Council, and with reference to what they have stated as to the inconvenience which may arise from the intended plan of limiting their Lordships' choice of members of the Resident Committee of Scientific Advice to the Council of the Royal Society, that their Lordships were induced to recommend this plan to his Majesty as a mark of respect to the Society, and as a pledge to the public of the qualification of the persons chosen. Nor did their Lordships apprehend any inconvenience from the circumstance stated in the Minute of the Council, of the Members being annually elected, as the Resident Committee is also annually appointed; and, in point of fact, no practical inconvenience has been felt during the ten years that the Committee has been in existence, as four of the distinguished gentlemen whom their Lordships have successively appointed to this office, have continued during the whole period to be members of the Council; and if any such difficulty or inconvenience should hereafter arise, their Lordships will be ready to take proper measures for remedying it.

Their Lordships' intention therefore is, to propose to Captain Kater and Mr. Herschel, to continue to fill this office; and to Dr. Young, who had resigned it, on receiving the appointment of Secretary to the late Board of Longitude, to be appointed.

I am, Sir, Your obedient servant,
JOHN BARROW.

The representation made by the Council was not calculated to produce much effect; but the Secretary of the Admiralty, who knew well the stuff of which Councils of the Royal Society are composed, might have spared the bitter irony of making their Lordships say, that they recommended this plan "AS A MARK OF RESPECT TO THE SOCIETY," and "AS A PLEDGE TO THE PUBLIC OF THE QUALIFICATIONS OF THE PERSONS CHOSEN," whilst he delicately hints to them their dependent situation, by observing, that the

"RESIDENT COMMITTEE IS ALSO ANNUALLY APPOINTED."

The Secretary knew that, PRACTICALLY speaking, it had been the custom for years for the President of the Royal Society to nominate the Council, and consequently he knew that every scientific adviser must first be indebted to the President for being qualified to advise, and then to the Admiralty for deriving profit from his counsel. Thus then their Lordships, as a "MARK OF RESPECT FOR THE SOCIETY" confirm the dependence of the Council on the President, by making his nomination a qualification for place, and establish a new dependence of the same Council on themselves, by giving a hundred pounds each year to such three members of that Council as they may select. "THE PLEDGE" they offer "TO THE PUBLIC, OF THE QUALIFICATIONS OF THE PERSONS CHOSEN," is, that Mr. Davies Gilbert had previously thought they would do for his Council.

What the Society, when they are acquainted with it, may think of this mark of respect, or what value the public may put upon this pledge, must be left to themselves to express.

In looking over the list of officers and Council of the Royal Society the weakest perhaps (for purposes of science) which was ever made, a consolation arises from the possibility of some of those who were placed there by way of compliment, occasionally attending. In that contracted field Lord Melville's penetration may not be uselessly employed; and the soldier who presides over our colonies may judge whether the principles which pervade it are open and liberal as his own.

The inconvenience to the public service from such an arrangement is, that the number out of which the advisers are selected must, in any case, be very small; and may, from several circumstances, be considerably reduced. In a council fairly selected, to judge of the merits of the various subjects likely to be brought under the consideration of the Society, anatomy, chemistry, and the different branches of natural history, will share with the numerous departments of physical science, in claiming to be represented by persons competently skilled in those subjects. These claims being satisfied, but few places will be left to fill up with mathematicians, astronomers, and persons conversant with nautical astronomy.

Let us look at the present Council. Is there a single mathematician amongst them, if we except Mr Barlow, whose deservedly high reputation rests chiefly on his physical and experimental inquiries, and whom the President and the Admiralty have clearly shown they do not look upon as a mathematician, by not appointing him an adviser?

Small as the number of those persons on the Council, who are conversant with the three subjects named in the Act of Parliament, must usually be, it may be still further diminished. The President, when he forms his Council, may decline naming those members who are most fit for such situations. Or, on the other hand, some of those members who are best qualified for them, from their knowledge, may decline the honour of being the nominees of Mr.

Gilbert, as Vice Presidents, Treasurers, or Councillors, and thus lending their names to support a system of which they disapprove.

Whether the first of these causes has ever operated can be best explained by those gentlemen who have been on the Council. The refusals are, notwithstanding the President's taciturnity on the subject, better known than he is willing that they should be.

Having discussed the general policy of the measure, with reference both to the Society and to the public, and without the slightest reference to the individuals who may have refused or accepted those situations, I shall now examine the propriety of the appointments that have been made.

Doubtless the gentlemen who now hold those situations either have never considered the influence such a mode of selection would have on the character of the Council; or, having considered it, they must have arrived at a different conclusion from mine. There may, however, be arguments which I have overlooked, and a discussion of them must ultimately lead to truth: but I confess that it appears to me the objections which have been stated rest on principles of human nature, too deeply seated to be easily removed.

That I am not singular in the view I have taken of this subject, appears from several circumstances. A question was asked respecting these appointments at the Anniversary before the last; and, from the nature of the answer, many of the members of the Society have been led to believe the objections have been removed. Several Fellows of the Society, who knew these facts, thought it inexpedient ever to vote for placing any gentleman on the Council who had accepted these situations; and, having myself the same view of the case, I applied to the Council to be informed of the names of the present Scientific Advisers. But although they remonstrated against the PRINCIPLE, they replied that they had "NO COGNIZANCE" of the fact.

The two first members of the Council, Mr. Herschel and Captain Kater, who were so appointed, and who had previously been Resident Commissioners under the Act, immediately refused the situations. Dr. Young became one of the Advisers; and Captain Sabine and Mr. Faraday were appointed by the Admiralty as the two remaining ones. Of Dr. Young, who died shortly after, I shall only observe that he possessed knowledge which qualified him for the situation.

Whether those who at present fill these offices can be said to belong to that class of persons which the Order in Council and the Act of Parliament point out, is a matter on which doubt may reasonably be entertained. The Order in Council speaks of these three persons as being the same, and having the "SAME DUTIES" as those mentioned in the Act; and it recites the words of the Act, that they shall be persons "WELL VERSED IN THE SCIENCES OF MATHEMATICS ASTRONOMY, AND NAVIGATION." Of the fitness of the gentlemen who now hold those situations to pronounce judgment on mathematical questions, the public will be better able to form an opinion when they shall have communicated to the world any of their own mathematical inquiries. Although it is the practice to consider that acceptance of office is alone necessary to qualify a man for a statesman, a similar doctrine has not yet prevailed in the world of science. One of these gentlemen, who has established his reputation as a chemist, stands in the same predicament with respect to the other two sciences. It remains then to consider Captain Sabine's claims, which must rest on his skill in "PRACTICAL ASTRONOMY AND NAVIGATION,"—a claim which can only be allowed when the scientific world are set at rest respecting the extraordinary nature of those observations contained in his work on the Pendulum.

That volume, printed under the authority of the Board of Longitude, excited at its appearance considerable attention. The circumstance of the Government providing instruments and means of transport for the purpose of these inquiries, placed at Captain Sabine's disposal means superior to those which amateurs can generally afford, whilst the industry with which he availed himself of these opportunities, enabled him to bring home multitudes of observations from situations rarely visited with such instruments, and for such purposes.

The remarkable agreement with each other, which was found to exist amongst each class of observations, was as unexpected by those most conversant with the respective processes, as it was creditable to one who had devoted but a few years to the subject, and who, in the course of those voyages, used some of the instruments for the first time in his life.

This accordance amongst the results was such,

that naval officers of the greatest experience, confessed themselves unable to take such lunars; whilst other observers, long versed in the use of the transit instrument, avowed their inability to take such transits. Those who were conversant with pendulums, were at a loss how to make, even under more favourable circumstances, similarly concordant observations. The same opinion prevailed on the continent as well as in England. On whatever subject Captain Sabine touched, the observations he published seemed by their accuracy to leave former observers at a distance. The methods of using the instruments scarcely differed in any important point from those before adopted; and, but for a fortunate discovery, which I shall presently relate, the world must have concluded that Captain Sabine possessed some keenness of vision, or acuteness of touch, which it would be hopeless for any to expect to rival.

The Council of the Royal Society spared no pains to stamp the accuracy of these observations with their testimony. They seem to have thrust Captain Sabine's name perpetually on their minutes, and in a manner which must have been almost distressing: they recommend him in a letter to the Admiralty, then in another to the Ordnance; and several of the same persons, in their other capacity, as members of the Board of Longitude, after voting him a THOUSAND POUNDS for these observations, are said to have again recommended him to the Master-General of the Ordnance. That an officer, commencing his scientific career, should be misled by such praises, was both natural and pardonable; but that the Council of the Royal

Society should adopt their opinion so heedlessly, and maintain it so pertinaciously, was as cruel to the observer as it was injurious to the interests of science.

It might have been imagined that such praises, together with the Copley medal, presented to Captain Sabine by the Royal Society, and the medal of Lalande, given to him by the Institute of France, had arisen from such a complete investigation of his observations, as should place them beyond the reach even of criticism. But, alas! the Royal Society may write, and nobody will attend; its medals have lost their lustre; and even the Institute of France may find that theirs cannot confer immortality. That learned body is in the habit of making most interesting and profound reports on any memoirs communicated to it; nothing escapes the penetration of their committees appointed for such purposes. Surely, when they enter on the much more important subject of the award of a medal, unusual pains must be taken with the previous report, and it might, perhaps, be of some advantage to science, and might furnish their admirers with arguments in their defence, if they would publish that on which the decree of their Lalande's medal to Captain Sabine was founded.

It is far from necessary to my present object, to state all that has been written and said respecting these pendulum experiments: I shall confine myself merely to two points; one, the transit observations, I shall allude to, because I may perhaps show the kind of feeling that exists respecting them, and possibly enable Captain Sabine to explain them. The other point, the error in the estimation of the division of the level, I shall discuss, because it is an admitted fact.

Some opinion may be formed of transit observations, by taking the difference of times of the passage of any star between the several wires; supposing the distances of those wires equal, the intervals of time occupied by the star in passing from one to the other, ought to be precisely the same. As those times of passing from one wire to another are usually given to seconds and tenths of seconds, it rarely happens that the accordance is perfect.

The transit instrument used by Captain Sabine was thirty inches in length, and the wires are stated to be equi-distant. Out of about 370 transits, there are eighty-seven, or nearly one-fourth, which have the intervals between all the wires agreeing to the same, the tenth of a second. At Sierra Leone, nineteen out of seventy-two have the same accordance; and of the moon culminating stars, p. 409, twelve out of twenty-four are equally exact. With larger instruments, and in great observatories, this is not always the case.

Captain Kater has given, in the Philosophical Transactions, 1819, p. 427, a series of transits, with a three and a half foot transit, in which about one-eleventh part of them only have this degree of accuracy; and it should be observed that not merely the instrument, but the stars selected, have, in this instance, an advantage over Captain Sabine's.

The transit of M. Bessel is five feet in length, made by Frauenhofer, and the magnifying power employed is 182; yet, out of some observations of his in January, 1826, only one-eleventh have this degree of accordance. In thirty-three of the Greenwich observations of January, 1828, fifteen have this agreement, or five-elevenths; but this is with a ten-feet transit. Now in none of these instances do the times agree within a tenth of a second between all the wires; but I have accounted those as agreeing in all the wires in which there is not more than four-tenths of a second between the greatest and least.

This superior accuracy of the small instrument requires some explanation. One which has been suggested is, that Captain Sabine employs a chronometer to observe transits with; and that since it beats five times in two seconds, each beat will give four-tenths of a second; and this being the smallest quantity registered, the agreement becomes more probable than if tenths were the smallest quantities noticed. In general, the larger the lowest unity employed the greater will be the apparent agreement amongst the differences. Thus, if, in the transit of stars near the pole, the times of passing the wires were only registered to the nearest minute, the intervals would almost certainly be equal. There is another circumstance, about which there is some difficulty. It is understood that the same instrument, —the thirty-inch transit, was employed by Lieutenant Foster; and it has not been stated that the wires were changed, although this has most probably been the case. Now, in the transits which the later observer has given, he has found it necessary to correct for a considerable inequality between the first and second wires (See Phil. Trans. 1827). If an erroneous impression has gone abroad on this subject, it is doing a service to science to insure its correction, by drawing attention to it.

Should these observations be confirmed by other observers, it would seem to follow that the use of a chronometer renders a transit more exact, and therefore that it ought to be used in observatories.

Among the instruments employed by Captain Sabine, was a repeating circle of six inches diameter, made by order of the Board of Longitude, for the express purpose of ascertaining how far repeating instruments might be diminished in size:—a most important subject, on which the Board seem to have entertained a very commendable degree of anxiety.

The following extract from the "Pendulum Experiments" is important:

"The repeating circle was made by the direction, and at the expense of the Board of Longitude, for the purpose of exemplifying the principle of repetition when applied to a circle of so small a diameter as six inches, carrying a telescope of seven inches focal length, and one inch aperture; and of practically ascertaining the degree of accuracy which might be retained, whilst the portability of the instrument should be increased, by a reduction in the size to half the amount which had been previously regarded by the most eminent artists as the extreme limit of diminution to which repeating circles, designed for astronomical purposes, ought to be carried.

"The practical value of the six-inch repeating circle may be estimated, by comparing the differences of the partial results from the mean at each station, with the correspondence of any similar collection of observations made with a circle, on the original construction, and of large dimensions; such, for instance, as the latitudes of the stations of the French are, recorded in the Base du Systeme Metrique: when, if due allowance be made for the extensive experience and great skill of the distinguished persons who conducted the French observations, the comparison will scarcely appear to the disadvantage of the smaller circle, even if extended generally through all the stations of the present volume; but if it be particularly directed to Maranham and Spitzbergen,—at which stations the partial results were more numerous than elsewhere, and obtained with especial regard to every circumstance by which their accuracy might be affected, the performance of the six-inch circle will appear fully equal to that of circles of the larger dimension. The comparison with the two stations, at which a more than usual attention was bestowed, is the more appropriate, because it was essential to the purposes for which the latitudes of the French stations were required, that the observations should always be conducted with the utmost possible regard to accuracy.

"It would appear, therefore, that in a repeating circle of six inches, the disadvantages of a smaller image enabling a less precise contact or bisection, and of an arch of less radius admitting of a less minute subdivision, may be compensated by the principle of repetition."

Captain Sabine has pointed out Maranham and Spitzbergen as places most favourable to the comparison. Let us take the former of these places, and compare the observations made there with the small repeating instrument of six inches diameter, with those made by the French astronomers at Formentera, with a repeating circle of forty-one centi-metres, or about sixteen inches in diameter, made by Fortin. It is singular that this instrument was directed, by the French Board of Longitude, to be made expressly for this survey, and the French astronomers paid particular attention to it, from the circumstance of some doubts having been entertained respecting the value of the principle of repetition.

The following series of observations were made with the two instruments. [I have chosen the inferior meridian altitude of Polaris, merely because the number of sets of observations are rather fewer. The difference between the extremes of the altitude of Polaris, deduced from sets taken above the pole by the same observers, amounts to seven seconds and a half.]

Latitude deduced from Polaris, with a repeating circle, 16 inches diameter.—BASE DU SYSTEME METRIQUE, tom. iv. p. 376. 1807.

Number of Observations.	Latitude of Formentera.	Names of Observers.
	deg. min. sec.	
64	38 39 55.3	Biot
100	54.7	Arago
10	56.2	Biot
88	56.9	Biot
120	56.7	Arago

84	54.9	Biot
100	56.5	Arago
102	57.1	Arago
80	54.5	Biot
88	53.3	Arago
90	53.6	Arago
88	53.8	Arago
92	53.7	Arago
42	55.6	Chaix
90	54.1	Chaix
80	53.9	Arago

Mean of 1318 Observations, 38deg. 39min. 54.93sec.

Sets of Observations made with a six-inch repeating circle, at Maranham.

Star.	Number of	Latitude	Observer.
	Observations.	deduced.	
		deg. min. sec.	
alpha Lyrae	8	2 31 42.4	Capt. Sabine
alpha Lyrae	12	43.8	Ditto
alpha Pavonis	10	44.5	Ditto
alpha Lyrae	12	44.6	Ditto
alpha Cygni	12	42.1	Ditto
alpha Gruris	12	42.2	Ditto

Mean latitude deduced from 66 observations 2deg. 31min 43.3sec.

In comparing these results, although the French observations were more than twenty times as numerous as the English, yet the deviations of the individual sets from the mean are greater. One second and three-tenths is the greatest deviation from the mean of the Maranham observations; whilst the greatest deviation of those of Formentera, is two seconds and two-tenths. If this mode of comparison should be thought unfair, on account of the greater number of the sets in the French observations, let any six, in succession, of those sets be taken, and compared with the six English sets; and it will be found that in no one instance is the greatest deviation from the mean of the whole of the observations less than in those of Maranham. It must also be borne in mind, that by the latitude deduced by the mean of 1250 superior culminations of Polaris by the same observers, the latitude of Formentera was found to be 38deg. 39min 57.07sec., a result differing by 2.14sec. from the mean of the 1318 inferior culminations given above. [This difference cannot be accounted for by any difference in the tables of refraction, as neither the employment of those of Bradley, of Piazzi, of the French, of Groombridge, of Young, of Ivory, of Bessel, or of Carlini, would make a difference of two-tenths of a second.]

These facts alone ought to have awakened the attention of Captain Sabine, and of those who examined and officially pronounced on the merits of his observations; for, supposing the skill of the observers equal, it seems a necessary consequence that "the performance of the six-inch circle is" not merely "fully equal to that of circles of larger dimensions," but that it is decidedly SUPERIOR to one of sixteen inches in diameter.

This opinion did indeed gain ground for a time; but, fortunately for astronomy, long after these observations were made, published, and rewarded, Captain Kater, having borrowed the same instrument, discovered that the divisions of its level, which Captain Sabine had considered to be equal to one second each, were, in fact, more nearly equal to eleven seconds, each one being 10.9sec. This circumstance rendered necessary a recalculation of all the observations made with that instrument: a recalculation which I am not aware Captain Sabine has ever thought it necessary to publish. [Above two hundred sets of observations with this instrument are given in the work alluded to. It can never be esteemed satisfactory merely to state the mean results of the corrections arising from this error: for the confidence to be attached to that mean will depend on the nature of the deviations from it.]

This is the more to be regretted, as it bears upon a point of considerable importance to navigation; and if it should have caused any alteration in his opinion as to the comparative merits of great and small instruments, it might have been expected from a gentleman, who was expressly directed by the Board of Longitude, to try the question with an instrument constructed for that especial purpose.

Finding that this has not been done by the person best qualified for the task, perhaps a few remarks from one who has no pretensions to familiarity with the instrument, may tend towards elucidating this interesting question.

The following table gives the latitudes as

corrected for the error of level:

Station.	Star	Latitude by Capt. Sabine	Latitude corrected for error of level.	Difference
		deg.min.sec.	deg.min.sec.	sec.
Sierra Leone	Sirius	8 29 27.9	8 29 34.7	6.8
Ascension	Alph.Centuri	7 55 46.7	7 55 40.1	6.6
Bahia	Alph.Lyrae	12 59 19.4	12 59 21.4	2.0
	Alph.Lyrae	21.2	58 49.8	31.4
	Alph.Pavonis	22.4	59 5.1	17.3
Maranham	Alph.Lyrae	2 31 42.4	2 31 22	20.4
	Alph.Lyrae	43.8	31.8	12.0
	Alph.Pavonis	44.5	44	.5
	Alph.Lyrae	44.6	42.6	2.0
	Alph.Cygni	42.1	39.2	2.9
	Alph.Gruris	42.2	27.4	14.8
Trinidad	Achernar	10 38 56.1	10 38 58.2	2.1

Alph.Gruris	52.2		50.8	1.4			
Achernar	59.3		56.6	2.7			

Jamaica	Polaris	17 56 8.6	17 56	4.6			
4.0							
		6.6		3.3	3.3		

New York	Sun	40 42 40.1	40 42				
44.6	4.5						
	Polaris	48.9		38.2	10.7		
	Sun	41.4		47.2	5.8		
	Beta Urs.Min.	42.3		58.4			
16.1							

Hammerfest	Sun	70 40 5.3	70 40				
7.2	1.9						

Spitzbergen	Sun	79 49 56.1	79 49				
58.6	2.5						
	Sun	55.9		44.8	11.1		
	Sun	58.6		52.7	5.9		
	Sun	59.3		51.6	7.7		
	Sun	55.8		51.6	4.2		
	Sun	50 1.5		57.0	4.5		

Greenland	Sun	74 32 19.9	74 32				
32.4	12.4						
	Sun	17.9		18.7	0.8		

Drontheim	Sun	63 25 51.3	63 26				
6.1	14.8						
	Alph.Urs.Min.	57.2		49.4			
7.8							

This presents a very different view of the latitudes as determined by the small repeating circle, from that in Captain Sabine's book; and confining ourselves still to Maranham, where the latitudes "WERE OBTAINED, WITH ESPECIAL REGARD TO EVERY CIRCUMSTANCE BY WHICH THEIR ACCURACY MIGHT BE AFFECTED," and where "A MORE THAN USUAL ATTENTION WAS BESTOWED," it appears, that if we take Captain Sabine's own test, namely, "the differences of the partial results from the mean at each station," the deviations become nearly ten times as large as they were before; a circumstance which might be expected to have some influence in the decision of the question.

There is, however, another light in which it is impossible to avoid looking at this singular oversight. The second column of the table of latitudes must now be considered the true one, as that which really resulted from the observations. Now, on examining the column of true latitudes, the differences between the different sets of observations is so considerable as naturally to excite some fear of latent error, more especially as nearly the greatest discordance arises from the same star, Alph.Lyrae, observed after an interval of only three days. It becomes interesting to every person engaged in making astronomical observations, to know what is the probability of his being exposed to an error so little to be guarded against, and so calculated to lull the suspicions of the unfortunate astronomer to whom it may happen.

In fact, the question resolves itself into this:

the true latitude of a place being determined by sets of observations as in the first of the following columns—

Latitudes as

True latitudes observed.   computed by a mistake

of Capt. Sabine's.

	deg.min.sec.	deg.min.sec.
Alph.Lyrae, 28th Aug....	2 31 22.0	2 31 42.4
Alph.Lyrae, 29th Aug....	31.8	43.8
Alph.Pavonis, 29th Aug...	44,0	44.5
Alph.Lyrae, 31st Aug....	42.6	44.6
Alph.Cygni, 31st Aug....	39.2	42.0
Alph.Gruris, 2d Sept....	27.4	42.2

what are the chances that, by one error all the latitudes in the first column should be brought so nearly to an agreement as they are in the second column? The circumstance of the number of divisions of the level being almost arbitrary within limits, might perhaps be alleged as diminishing this extraordinary improbability: but let any one consider, if he choose the error of each set, as independent of the others, still he will find the odds against it enormous.

When it is considered that an error, almost arbitrary in its law, has thus had the effect of bringing discordant observations into an almost unprecedented accordance, as at Maranham; and not merely so, but that at eight of the nine stations it has uniformly tended to diminish the differences between the partial results, and that at the ninth station it only increased it by a small fraction of a second, I cannot help feeling that it is more probable even that Captain Kater, with all his admitted skill, and that Captain Sabine himself, should have been both mistaken in their measures of the divisions of the level, than that so singular an effect should have been produced by one error; and I cannot bring myself to believe that such an anticipation is entirely without foundation.

Whatever may be the result of a re-examination, it was a singular oversight NOT TO MEASURE the divisions of a level intended to be used for determining so important a question; more particularly as, in the very work to which reference was made by Captain Sabine for the purpose of comparing the observations, it was the very first circumstance which occupied the French philosophers, and several pages [See pages 265 to 275 of the RECUEIL D'OBSERVATIONS GEODESIQUES, &c. PAR MM. BIOT ET ARAGO, which forms the fourth volume of the BASE DU SYSTEME METRIQUE.] are filled with the details relative to the determination of the value of the divisions of the level. It would also have been satisfactory, with such an important object in view, to have read off some of the sets after each pair of observations, in order to see how far the system of repetition made the results gradually converge to a limit, and in order to know how many repetitions were sufficient. Such a course would almost certainly have led to a knowledge of the true value of the divisions of the level; for the differences in the altitude of the same star, after a few minutes of time, must, in many instances, have been far too great to have arisen from the change of its altitude: and had these been noticed, they must have been referred to some error in the instrument, which could scarcely, in such circumstances, have escaped detection.

I have now mentioned a few of the difficulties which attend Captain Sabine's book on the pendulum, difficulties which I am far from saying are inexplicable. He would be bold indeed who, after so wonderful an instance of the effect of chance as I have been just discussing, should venture to pronounce another such accident impossible; but I think enough has been said to show, that the feeling which so generally prevails relative to it, is neither captious nor unreasonable.

Enough also has appeared to prove, that the conduct of the Admiralty in appointing that gentleman one of their scientific advisers, was, under the peculiar circumstances, at least, unadvised. They have thus lent, as far as they could, the weight of their authority to support observations which are now found to be erroneous. They have thus held up for imitation observations which may induce hundreds of meritorious officers to throw aside their instruments, in the despair of ever approaching a standard which is since admitted to be imaginary; and they have ratified the doctrine, for I am not aware their official adviser has ever even modified it, that diminutive instruments are equal almost to the largest.

To what extent this doctrine is correct, may perhaps yet admit of doubt. It cannot, however, admit of a doubt, that it is unwise to crown it with official authority, and thus expose the officers of their service to depend on means which may be quite insufficient for their purpose.

How the Board of Longitude, after EXPRESSLY DIRECTING THIS INSTRUMENT TO BE MADE AND TRIED, could come to the decision at which they arrived, appears inexplicable. The known difference of opinion amongst the best observers respecting the repeating principle, ought to have rendered them peculiarly cautious, nor ought the opinion of a Troughton, that instruments of less than one foot in diameter may be considered, "FOR ASTRONOMY, AS LITTLE BETTER THAN PLAYTHINGS," [Memoirs of the Astronomical Society, Vol.I. p.53.] to have been rejected without the most carefully detailed experiments. There were amongst that body, persons who must have examined minutely the work on the Pendulum. Captain Kater must have felt those difficulties in the perusal of it which other observers have experienced; and he who was placed in the Board of Longitude especially for his knowledge of instruments, might, in a few hours, have arrived at more decisive facts. But perhaps I am unjust. Captain Kater's knowledge rendered it impossible for him to have been ignorant of the difficulties, and his candour would have prevented him from concealing them: he must, therefore, after examining the subject, have been outvoted by his lay-brethren who had dispensed with that preliminary.

It would be unjust, before quitting this subject, not to mention with respect the acknowledgment made by an officer of the naval service of the errors into which he also fell from this same level. Lieutenant Foster, aware of the many occasions on which Captain Sabine had employed this instrument, and knowing that he considered each division as equal to one second, never thought that a doubt could exist on the subject, and made all his calculations accordingly. When Captain Kater made him acquainted with the mistake, Lieutenant Foster immediately communicated a paper [The paper of Lieutenant Foster is printed in the Philosophical Transactions, 1827, p.122, and is worth consulting.] to the Royal Society, in which he states the circumstance most fully, and recomputed all the observations in which that instrument was used. Unfortunately, from the original observations of Mr. Ross being left on board the Fury at the time of her loss, the transcripts of his results could not be recomputed like the rest, and were consequently useless.

SECTION 5. OF THE UNION OF SEVERAL OFFICES IN ONE PERSON.

Although the number of situations to which persons conversant with science may hope to be appointed, is small, yet it has somewhat singularly happened, that instances of one individual, holding more than one such appointment, are frequent. Not to speak of those held by the late Dr. Young, we have at present:—

MR. POND—Astronomer Royal, Inspector of Chronometers, and Superintendent of the Nautical Almanac.

CAPTAIN SABINE—An officer of artillery on leave of absence from his regiment; Secretary of the Royal Society; and Scientific Adviser of the Admiralty.

MR. BRANDE—Clerk of the Irons at the Royal Mint; Professor of Chemistry at the Royal Institution; Analyser of Rough Nitre, &c. to the East-India Company; Lecturer on Materia Medica, Apothecaries' Hall; Superintending Chemical Operator at ditto; Lecturer on Chemistry at ditto; Editor of the Royal Institution Journal; and Foreign Secretary to the Royal Society.

One should be led to imagine, from these unions of scientific offices, either that science is too little paid, and that gentlemen cannot be found to execute the offices separately at the salaries offered; or else, that it is too well paid, since each requires such little attention, that almost any number can be executed by one person.

The Director of the Royal Observatory has a larger and better collection of instruments, and more assistants to superintend, than any other astronomer in the world; and, to do it properly, would require the almost undivided attention of a man in the vigour of youth. Nor would a superintendent of the Nautical Almanac, if he made a point of being acquainted with every thing connected with his subject, find his situation at all a sinecure. Slight as are the duties of the Foreign Secretary of the Royal Society, it might have been supposed that Mr. Brande would scarcely, amongst his multifarious avocations, have found time even for them. But it may be a consolation to him to know, that from the progress the Society is making, those duties must become shortly, if they are not already, almost extinct.

Doubtless the President, in making that appointment, looked most anxiously over the list of the Royal Society. He doubtless knew that the Academics of Sweden, of Denmark, of Scotland, of Prussia, of Hanover, and of France, derived honour from the discoveries of their Secretaries;—that they prided themselves in the names of Berzelius, of Oersted, of Brewster, of Encke, of Gauss, and of Cuvier. Doubtless the President must have been ambitious that England should contribute to this galaxy of glory, that the Royal Society should restore the lost Pleiad [Pleiades, an assemblage of seven stars in the neck of the constellation Taurus. There are now only six of them visible to the naked eye.—HUTTON'S DICTIONARY—Art. Pleiades.] to the admiring science of Europe. But he could discover no kindred name amongst the ranks of his supporters, and forgot, for a moment, the interest of the Society, in an amiable consideration for the feelings of his surrounding friends. For had the President chosen a brighter star, the lustre of his other officers might have been overpowered by its splendour: but relieved from the pain of such a contrast, he may still retain the hope, that, by their united brightness, these suns of his little system shall yet afford sufficient light to be together visible to distant nations, as a faint NEBULA in the obscure horizon of English science.

SECTION 6. OF THE FUNDS OF THE SOCIETY.

Although the Society is not in a state approaching to poverty, it may be useful to offer a few remarks respecting the distribution of its money.

EXPENSE OF ENGRAVINGS FOR SIR E. HOME'S PAPERS.—The great expense of the engravings which adorn the volumes of the Philosophical Transactions, is not sufficiently known. That many of those engravings are quite essential for the papers they illustrate, and that those papers are fit for the Transactions, I do not doubt; but, some inquiry is necessary, when such large sums are expended. I shall endeavour, therefore, to approximate to the sum these engravings have cost the Royal Society.

Previous to 1810, there are upwards of seventy plates to papers of Sir E. Home's; in many of these, which I have purposely separated, the workmanship is not so minute as in the succeeding ones. Since 1810, there have occurred 187 plates attached to papers of the same author. Many of these have cost from twelve to twenty guineas each plate; but I shall take five pounds as the average cost of the first portion, and twelve as that of the latter. This would produce,

70 X 5 = 350
187 X 12 = 2244
...... ————-...... L2594

As this is only proposed as a rough approximation, let us omit the odd hundreds, and we have two thousand pounds expended in plates only on ONE branch of science, and for one person! Without calling in question the importance of the discoveries contained in those papers, it may be permitted to doubt whether such a large sum might not have been expended in a manner more beneficial to science. Not being myself conversant with those subjects, I can only form an opinion of the value from extraneous circumstances. Had their importance been at all equal to their number, I should have expected to have heard amongst the learned of other countries much more frequent mention of them than I have done, and even the Council of the Royal Society would scarcely have excluded from their Transactions one of those productions which they had paid for as a lecture.

It might also have been more delicate not to have placed on the Council so repeatedly a gentleman, for whose engravings they were annually expending, during the last twenty years, about an hundred pounds. On the other hand, when the Council lent Sir E. Home the whole of those valuable plates to take off impressions for his large work on Comparative Anatomy, of which they constitute almost the whole, it might have been as well not to have obliterated from each plate all indication of the source to which he was indebted for them.

THE PRESIDENT'S DISCOURSES.—I shall mention this circumstance, because it fell under my own observation.

Observing in the annual accounts a charge of 381L 5s. for the President's Speeches, I thought it right to inquire into the nature of this item. Happening to be on the Council the next year, I took an opportunity, at an early meeting of that Council, to ask publicly for an explanation of the following resolution, which stands in the Council-books for Dec. 21, 1828.

"Resolved, That 500 copies of the President's Discourses, about to be printed by Mr. Murray, be purchased by the Society, at the usual trade price."

The answer given to that question was, "THAT THE COUNCIL HAD AGREED TO PURCHASE THESE VOLUMES AT THAT PRICE, IN ORDER TO INDUCE MR. MURRAY TO PRINT THE PRESIDENT'S SPEECHES."

I remarked at the time that such an answer was quite unsatisfactory, as the following statement will prove.

The volume consists of 160 pages, or twenty sheets, and the following prices are very liberal:

	L	s.	d.
To composing and printing twenty sheets, at 3L. per sheet...............	60	0	0
Twenty reams of paper, at 3L. per ream.....	60	0	0

| Corrections, alterations, &c.......... | 30 | 0 | 0 |
| Total cost of 500 copies...... | 150 | 0 | 0 |

Now upon the subject of the expense of printing, the Council could not plead ignorance. The Society are engaged in printing, and in paying printers' bills, too frequently to admit of such an excuse; and several of the individual members must have known, from their own private experience, that the cost of printing such a volume was widely different from that they were about to pay, as an inducement to a bookseller to print it on his own account. Here, then, was a sum of above two hundred pounds beyond what was necessary for the object, taken from the funds of the Royal Society; and for what purpose? Did the President and his officers ever condescend to explain this transaction to the Council; or were they expected, as a matter of course, to sanction any thing proposed to them? Could they have been so weak, or so obedient, as to order the payment of above three hundred and eighty pounds, to induce a bookseller to do what they might have done themselves for less than half the sum? Or did they wish to make Mr. Murray a present of two hundred pounds? If so, he must have had powerful friends in the Council, and it is fit the Society should know who they were; for they were not friends, either to its interests or to its honour.

The copies, so purchased, were ordered by the Council to be sold to members of the Society at 15s. each: (the trade price is 15s. 3d.) and out of the five hundred copies twenty-seven only have been sold: the remainder encumber our shelves. Thus, after four years, the Society are still losers of three hundred and sixty Pounds on this transaction.

ON THE CONVERSION OF THE GREENWICH OBSERVATIONS INTO PASTEBOARD.—Although the printing of these observations is not paid for out of the funds of the Royal Society, yet as the Council of that body are the visitors of the Royal Observatory, it may not be misplaced to introduce the subject here.

Some years since, a member of the Royal Society accidentally learned, that there was, at an old store-shop in Thames Street, a large quantity of the volumes of the Greenwich Observations on sale as waste paper. On making inquiry, he ascertained that there were two tons and a half to be disposed of, and that an equal quantity had already been sold, for the purpose of converting it into pasteboard. The vendor said he could get fourpence a pound for the whole, and that it made capital Bristol board. The fact was mentioned by a member of the Council of the Royal Society, and they thought it necessary to inquire into the circumstances.

Now, the Observations made at the Royal Observatory are printed with every regard to typographical luxury, with large margins, on thick paper, hotpressed, and with no sort of regard to economy. This magnificence is advocated by some who maintain, that the volumes ought to be worthy of a great nation; whilst others, seeing how little that nation spends on science, regret that the sums allotted to it should not be applied with the strictest economy. If the Astronomer Royal really has a right to these volumes, printed by the government at a large expense, it is, perhaps, the most extravagant mode which was ever yet invented of paying a public servant. When that right was given to him,— let us suppose somebody had suggested the impolicy of it, lest he should sell the costly volumes for waste paper,—who would have listened for one moment to such a supposition? He would have been told that it was impossible to suppose a person in that high and responsible situation, could be so indifferent to his own reputation.

A short time since, I applied to the President and Council of the Royal Society, for copies of the Greenwich Observations, which were necessary for an inquiry on which I was at that time engaged. Being naturally anxious to economize the small funds I can devote to science, the request appeared to me a reasonable one. It was, however, refused; and I was at the same time informed that the Observations could be purchased at the bookseller's. [This was a mistake; Mr. Murray has not copies of the Greenwich Observations prior to 1823.] When I consider that practical astronomy has not occupied a very prominent place in my pursuits, I feel disposed, on that ground, to acquiesce in the propriety of the refusal. This excuse can, however, be of no avail for similar refusals to other gentlemen, who applied nearly at the same time with myself, and whose time had been successfully devoted to the cultivation of that science. [M. Bessel, at the wish of the Royal Academy of Berlin, projected a plan for making a very extensive map of the heavens. Too vast for any individual to attempt, it was proposed that a portion should be executed by the astronomers of various countries, and invitations to this effect were widely circulated. One only of the divisions of this map was applied for by any English astronomer; and, after completing the portion of the map assigned to him, he undertook another, which had remained unprovided for. This gentleman, the Rev. Mr. Hussey, was one of the rejected applicants for the Greenwich Observations.]

There was, however, another ground on which I had weakly anticipated a different result;—but those who occupy official situations, rendered remarkable by the illustrious names of their predecessors, are placed in no enviable station; and, if their own acquirements are confessedly insufficient to keep up the high authority of their office, they must submit to the mortifications of their false position. I am sure, therefore, that the President and officers of the Royal Society must have sympathized MOST DEEPLY with me, when they felt it their duty to propose that the Society over which Newton once presided, should refuse so trifling an assistance to the unworthy possessor of the chair he once filled.

In reply to my application to the President and Council, to be allowed a copy of the Greenwich Observations, I was informed that, "The number of copies placed by government at the disposal of the Royal Society, was insufficient to supply the demands made on them by various learned bodies in Europe; and, consequently, they were unable, however great their inclination, to satisfy the wishes of individual applicants." Now I have spent some time in searching the numerous proceedings in the council-books of the Royal Society, and I believe the following is the real state of the case:—

In 1785, Lord Sidney, one of His Majesty's principal Secretaries of State, wrote to the Council a letter, dated Whitehall, March 8, 1785, from which the following is extracted:—

"The King has been pleased to consent, that any copies of the Astronomical Observations, made at the Observatory of Greenwich, (and paid for by the Board of Ordnance, pursuant to His Majesty's command, of July 21, 1767,) which may at any time remain in the hands of the printer, shall, after you have reserved such copies as you may think proper as presents, be given to the said Nevil Maskelyne, in consideration of his trouble in the superintending the printing thereof. I am to signify His Majesty's pleasure, that you do, from time to time, give the necessary orders for that purpose, until His Majesty's further commands shall be communicated to you.

Soon after this letter, I find on the council-books:—

"Ordered, That sixty copies of the Greenwich Observations, last published, be retained as presents, and that the rest be delivered to the Astronomer Royal."

It is difficult to be sure of a negative fact, but in searching many volumes of the Proceedings of the Council, I have not discovered any revocation of this order, and I believe none exists. This is confirmed by the circumstance of the Council at the present day receiving precisely the same number of copies as their predecessors, and I believe that in fact they do not know the authority on which the right to those sixty rests.

Supposing this order unrevoked, it was clearly meant to be left to the discretion of the Council, to order such a number to be reserved, "from time to time," as the demands of science might require. When, therefore, they found that the number of sixty copies was insufficient, they ought to have directed the printer to send them a larger number; but when they found out the purpose to which the Astronomer Royal applied them, they ought immediately to have ordered nearly the whole impression, in order to prevent this destruction of public property. If, on the other hand, the above order is revoked, and we really have no right to more than sixty copies; then, on discovering the Observations in their progress towards pasteboard, it was the duty of the Council of the Royal Society, as visitors of the Royal Observatory, immediately to have represented to Government the evil of the arrangement, and to have suggested, that if the Astronomer Royal have the right, it would be expedient to commute it for a liberal compensation.

Whichever be the true view of the case, they have taken no steps on the subject; and I cannot help expressing my belief, that the President and Council were induced to be thus negligent of the interests of science, from the fear of interfering with the perquisites of the Astronomer Royal.

It is, however, but justice to observe, that the injury already done to science, by the conversion of these Observations into pasteboard, is not so great as the public might have feared. Mr. Pond, than whom no one can be supposed better acquainted with their value, and whose right to judge no man can question, has shown his own opinion to be, that his reputation will be best consulted by diminishing the extent of their circulation.

Before I quit the subject of the Royal Observatory, on which much might be said, I will just refer to the report by a Committee of the Royal Society that was made relative to it, some years since, and which, it is imagined, is a subject by no means grateful to the memory of any of the parties concerned in it. My object is to ascertain, whether any amendments have taken place in consequence. To one fact of considerable importance, I was myself a witness, when I was present officially at a visitation. At that time, no original observations made at the transit instrument were ever preserved. Had I not been an eye witness of the process of an observation, I should not have credited the fact.

SECTION 7. OF THE ROYAL MEDALS.

At a period when the attention of Government to science had not undergone any marked change, a most unexpected occurrence took place. His Majesty intimated to the Royal Society, through his Secretary of State, his intention to found two gold medals, of the value of fifty guineas each, to be awarded annually by the Council of the Royal Society, according to the rules they were desired to frame for that purpose.

The following is the copy of Mr. Peel's letter:—

WHITEHALL, December 3d, 1825.

SIR,

I am commanded by the King to acquaint you, that His Majesty proposes to found two gold medals, of the value of fifty guineas each, to be awarded as honorary premiums, under the direction of the President and Council of the Royal Society, in such a manner as shall, by the excitement of competition among men of science, seem best calculated to promote the object for which the Royal Society was instituted.

His Majesty desires to receive from the President and Council of the Royal Society their opinion upon the subject generally of the regulations which it may be convenient to establish with regard to the appropriation of the medals; and I have, therefore, to request that you will make the necessary communication to the Council of the Royal Society, in order that His Majesty's wishes may be carried into effect.

I have the honour to be, &c. &c. (Signed) R. PEEL.

Nothing could be more important for the interests of science, than this gracious manifestation of His Majesty's concern for its advancement. It was hailed by all who were made acquainted with it, as the commencement of a new era, and the energies which it might have awakened were immense. The unfettered nature of the gift excited admiration, whilst the confidence reposed in the Council was calculated to have insured the wavering faith of any less-gifted body. Even those who, either from knowing the MANAGEMENT of the Society, or from other grounds, doubted the policy of establishing medals, saw much to admire in the tone and spirit in which they were offered.

The Council immediately came to the resolution of gratefully accepting them: and it appears that the President communicated that resolution, on the 26th, to Mr. Peel, in a letter, which is found on the minutes of the Council-book of the 26th of January.

At the same Council, the rules for the award of the Royal medals were decided upon; they were as follow:—

26th January, 1826.

RESOLVED,

That it is the opinion of the Council, that the medals be awarded for the most important discoveries or series of investigations, completed and made known to the Royal Society in the year preceding the day of their award.

That it is the opinion of the Council, that the presentation of the medals should not be limited to British subjects. And they propose, if it should be His Majesty's pleasure, that his effigy should form the obverse of the medal.

That two medals from the same die should be struck upon each foundation; one in gold, one in silver.

If these rules are not the wisest which might

have been formed, yet they are tolerably explicit; and it might have been imagined that even a councillor of the Royal Society, prepared for office by the education of a pleader, could not have mystified his brethren so completely, as to have made them doubt on the point of time. The rules fixed precisely, that the discoveries or experiments rewarded, must be completed and made known to the Royal Society, within the YEAR PRECEDING THE DAY of the award.

Perhaps it might have been a proper mark of respect to this communication, to have convened a special general meeting of the Society, to have made known to the whole body the munificent endowment of their Patron: and when his approbation of the laws which were to govern the distribution of these medals had been intimated to the Council, such a course would have been in complete accordance with the wish expressed in Mr. Peel's letter, "TO EXCITE COMPETITION AMONGST MEN OF SCIENCE" by making them generally known.

Let us now examine the first award of these medals: it is recorded in the following words:—

November 16, 1826.

ONE of the medals of His Majesty's donation for the present year was awarded to John Dalton, Esq. President of the Philosophical and Literary Society, Manchester, for his development of the Atomic Theory, and his other important labours and discoveries in physical science.

The other medal for the present year was awarded to James Ivory, Esq. for his paper on Astronomical Refractions, published in the

Philosophical Transactions for the year 1823, and his other valuable papers on mathematical subjects.

The Copley medal was awarded to James South, Esq. for his observations of double stars, and his paper on the discordances between the sun's observed and computed right ascensions, published in the Transactions.

It is difficult to believe that the same Council, which, in January, formed the laws for the distribution of these medals, should meet together in November, and in direct violation of these laws, award them to two philosophers, one of whom had made, and fully established, his great discovery almost twenty years before; and the other of whom (to stultify themselves still more effectually) they expressly rewarded for a paper made known to them three years before.

Were the rules for the award of these medals read previous to their decision? Or were the obedient Council only used to register the edict of their President? Or were they mocked, as they have been in other instances, with the semblance of a free discussion?

Has it never occurred to gentlemen who have been thus situated, that although they have in truth had no part in the decision, yet the Society and the public will justly attribute a portion of the merit or demerit of their award, to those to whom that trust was confided?

Did no one member of the Council venture, with the most submissive deference, to suggest to the President, that the public eye would watch with interest this first decision on the Royal medals, and that it might perhaps be more discreet to adjudge them, for the first time, in accordance with the laws which had been made for their distribution? Or was public opinion then held in supreme contempt? Was it scouted, as I have myself heard it scouted, in the councils of the Royal Society?

Or was the President exempt, on this occasion, from the responsibility of dictating an award in direct violation of the faith which had been pledged to the Society and to the public? and, did the Council, intent on exercising a power so rarely committed to them; and, perhaps, urged by the near approach of their hour of dinner, dispense with the formality of reading the laws on which they were about to act?

Whatever may have been the cause, the result was most calamitous to the Society. Its decision was attacked on other grounds; for, with a strange neglect, the Council had taken no pains to make known, either to the Society, or to the public, the rules they had made for the adjudication of these medals.

The evils resulting from this decision were many. In the first place, it was most indecorous and ungrateful to treat with such neglect the rules which had been approved by our Royal Patron. In the next place, the medals themselves became almost worthless from this original taint: and they ceased to excite "competition amongst men of science," because no man could feel the least security that he should get them, even though his discoveries should fulfil all the conditions on which they were offered,

The great injury which accrued to science from

this proceeding, induced me, in the succeeding session, when I found myself on the Council of the Royal Society, to endeavour to remove the stigma which rested on our character. Whether I took the best means to remedy the evil is now a matter of comparatively little consequence: had I found any serious disposition to set it right, I should readily have aided in any plans for doing that which I felt myself bound to attempt, even though I should stand alone, as I had the misfortune of doing on that occasion. [It is but justice to Mr. South, who was a member of that Council, to state, that the circumstance of his having had the Copley medal of the same year awarded to him, prevented him from taking any part in the discussion.]

The impression which the whole of that discussion made on my mind will never be effaced. Regarding the original rules formed for the distribution of the Royal medals, when approved by his Majesty, as equally binding in honour and in justice, I viewed the decision of the Council, which assigned those medals to Mr. Dalton and Mr. Ivory, as void, IPSO FACTO, on the ground that it was directly at variance with that part which CONFINES the medals to discoveries made known to the Society within ONE YEAR PREVIOUS TO THE DAY OF THEIR AWARD. I therefore moved the following resolutions:

"1st, That the award of the Royal medals, made on the 16th of November, 1826, being contrary to the conditions under which they were offered, is invalid.

"2dly, That the sum of fifty guineas each be

presented to J. Dalton, Esq. and James Ivory, Esq. from the funds of the Society; and that letters be written to each of those gentlemen, expressing the hope of the Council that this, the only method which is open to them of honourably fulfilling their pledges, will be received by those gentlemen as a mark of the high sense entertained by the Council of the importance and value of their discoveries, which require not the aid of medals to convey their reputation to posterity, as amongst the greatest which distinguished the age in which they lived."

It may be curious to give the public a specimen of the reasoning employed in so select a body of philosophers as the Council of the Royal Society. It was contended, on the one hand, that although the award was SOMEWHAT IRREGULAR, yet nothing was more easy than to set it right. As the original rules for giving the medals were merely an order of the Council,—it would only be necessary to alter them, and then the award would agree perfectly with the laws. On the other hand, it was contended, that the original rules were unknown to the public and to the Society; and that, in fact, they were only known to the members of the Council and a few of their friends; and therefore the award was no breach of faith.

All comment on such reasoning is needless. That such propositions could not merely be offered, but could pass unreproved, is sufficient to show that the feelings of that body do not harmonize with those of the age; and furnishes some explanation why several of the most active members of the Royal Society have declined connecting their names with the Council as long as the present system of management is pursued.

The little interest taken by the body of the Society, either in its peculiar pursuits, or in the proceedings of the Council, and the little communication which exists between them, is an evil. Thus it happens that the deeds of the Council are rarely known to the body of the Society, and, indeed, scarcely extend beyond that small portion who frequent the weekly meetings. These pages will perhaps afford the first notice to the great majority of the Society of a breach of faith by their Council, which it is impossible to suppose a body, consisting of more than six hundred gentlemen, could have sanctioned.

SECTION 8. OF THE COPLEY MEDALS.

An important distinction exists between scientific communications, which seems to have escaped the notice of the Councils of the Royal Society. They may contain discoveries of new principles,—of laws of nature hitherto unobserved; or they may consist of a register of observations of known phenomena, made under new circumstances, or in new and peculiar situations on the face of our planet. Both these species of additions to our knowledge are important; but their value and their rarity are very different in degree. To make and to repeat observations, even with those trifling alterations, which it is the fashion in our country (in the present day) to dignify with the name of discoveries, requires merely inflexible candour in recording precisely the facts which nature has presented, and a power of fixing the attention on the instruments employed, or phenomena examined,—a talent, which can be much improved by proper Instruction, and which is possessed by most persons of tolerable abilities and education.* To discover new principles, and to detect the undiscovered laws by which nature operates, is another and a higher task, and requires intellectual qualifications of a very different order: the labour of the one is like that of the computer of an almanac; the inquiries of the other resemble more the researches of the accomplished analyst, who has invented the formula: by which those computations are performed.

[*That the use even of the large astronomical instruments in a national observatory, does not require any very profound acquirements, is not an opinion which I should have put forth without authority. The Astronomer-Royal ought to be the best judge.

On the minutes of the Council of the Royal Society, for April 6, 1826, with reference to the Assistants necessary for the two mural circles, we find a letter from Mr. Pond on the subject, from which the following passage is extracted:

"But to carry on such investigations, I want indefatigable, hard-working, and above all, obedient

120

drudges (for so I must call them, although they are drudges of a superior order), men who will be contented to pass half their day in using their hands and eyes in the mechanical act of observing, and the remainder of it in the dull process of calculation."]

Such being the distinction between the merits of these inquiries, some difference ought to exist in the nature of any rewards that may be proposed for their encouragement. The Royal Society have never marked this difference, and consequently those honorary medals which are given to observations, gain a value which is due to those that are given for discoveries; whilst these latter are diminished in their estimation by such an association.

I have stated this distinction, because I think it a just one; but the public would have little cause of complaint if this were the only ground of objection to the mode of appropriating the Society's medals. The first objection to be noticed, is the indistinct manner in which the object for which the medals are awarded is sometimes specified. A medal is given to A. B. "for his various papers."

There are cases, few perhaps in number, where such a reason may be admissible; but it is impossible not to perceive the weakness of those who judge these matters legibly written in the phrase, "and for his various other communications," which comes in as the frequent tail-piece to these awards. With a diffidence in their own powers, which might be more admired if it were more frequently expressed, the Council think to escape through this loop-hole, should the propriety of their judgment on the main point be called in question. Thus, even the discovery which made chemistry a science, has attached to it in their award this feeble appendage.

It has been objected to the Royal Society, that their medals have been too much confined to a certain set. When the Royal medals were added to their patronage, the past distribution of the Copley medals, furnished grounds to some of the journals to predict the future possessors of the new ones. I shall, doubtless, be told that the Council of the Royal Society are persons of such high feeling, that it is impossible to suppose their decision could be influenced by any personal motives. As I may not have had sufficient opportunities, during the short time I was a member of that Council, to enable me to form a fair estimate, I shall avail myself of the judgment of one, from whom no one will be inclined to appeal, who knew it long and intimately, and who expressed his opinion deliberately and solemnly.

The late Dr. Wollaston attached, as a condition to be observed in the distribution of the interest of his munificent gift of 2,000L. to the Royal Society, the following clause:—"And I hereby empower the said President, Council, and Fellows, after my decease, in furtherance of the above declared objects of the trust, to apply the said dividends to aid or reward any individual or individuals of any country, SAVING ONLY THAT NO PERSON BEING A MEMBER OF THE COUNCIL FOR THE TIME BEING, SHALL RECEIVE OR PARTAKE OF SUCH REWARD."

Another improvement which might be suggested, is, that it is generally inexpedient to vote a medal until the paper which contains the discovery is at least read to the Society; perhaps even it might not be quite unreasonable to wish that it should have been printed, and consequently have been perused by some few of those who have to decide on its merits. These trifles have not always been attended to; and even so lately as the last year, they escaped the notice of the President and his Council. The Society was, however, indebted to the good sense of Mr. Faraday, who declined the proffered medal; and thus relieved us from one additional charge of precipitancy. [When this hasty adjudication was thus put a stop to, one of the members of the Council inquired, whether, as a Copley medal must by the will be annually given, some other person might not be found deserving of it. To which the Secretary replied, "We do not intend to give any this year." All further discussion was thus silenced.]

Perhaps, also, as the Council are on some occasions apt to be oblivious, it might be convenient that the President should read, previously to the award of any medals or to the decision of any other important subjects, the statutes relating to them. He might perhaps propitiate their attention to them, by stating, HOW MUCH IT IMPORTETH TO THE CONSISTENCY OF THE COUNCIL TO BE ACQUAINTED WITH THE LAWS ON WHICH THEY ARE ABOUT TO DECIDE.

If those who have been conversant with the internal management of the Council, would communicate their information, something curious might perhaps be learned respecting a few of these medals. Concerning those of which I have had good means of information, I shall merely state—of three

of them—that whatever may have been the official reasons for their award, I had ample reasons to convince me of the following being the true causes:
—

First.—A medal was given to A, at a peculiarly inappropriate time—BECAUSE HE HAD NOT HAD ONE BEFORE.

Second.—Subsequently a medal was given to B, in order TO DESTROY THE IMPRESSION WHICH THE AWARD OF THE MEDAL TO A HAD MADE ON THE PUBLIC THE PRECEDING YEAR.

Third.—A medal was given to C, "BECAUSE WE THINK HE HAS BEEN ILL USED."

I will now enter on an examination of one of their awards, which was peculiarly injudicious. I allude to that concerning the mode of rendering platina malleable. Respecting, as I did, the illustrious philosopher who invented the art, and who has left many other claims to the gratitude of mankind, I esteem it no disrespect to his memory to place that subject in its proper light.

An invention in science or in art, may justly be considered as possessing the rights of property in the highest degree. The lands we inherit from our fathers, were cultivated ere they were born, and yielded produce before they were cultivated. The products of genius are the actual creations of the individual; and, after yielding profit or honour to him, they remain the permanent endowments of the human race. If the institutions of our country, and the opinions of society, support us fully in the absolute disposal of our fields, of which we can, by the laws of nature, be only the transitory possessors, who shall justly restrict our discretion in the disposal of those richer possessions, the products of intellectual exertion?

Two courses are open to those individuals who are thus endowed with Nature's wealth. They may lock up in their own bosoms the mysteries they have penetrated, and by applying their knowledge to the production of some substance in demand in commerce, thus minister to the wants or comforts of their species, whilst they reap in pecuniary profit the legitimate reward of their exertions.

It is open to them, on the other hand, to disclose the secret they have torn from Nature, and by allowing mankind to participate with them, to claim at once that splendid reputation which is rarely refused to the inventors of valuable discoveries in the arts of life.

The two courses are rarely compatible, only indeed when the discoverer, having published his process, enters into equal competition with other manufacturers.

If an individual adopt the first of these courses, and retaining his secret, it perish with him, the world have no right to complain. During his life, they profited by his knowledge, and are better off than if the philosopher had not existed.

Monopolies, under the name of patents, have been devised to assist and reward those who have chosen the line of pecuniary profit. Honorary rewards and medals have been the feeble expressions of the sentiments of mankind towards those who have preferred the other course. But these have been, and should always be, kept completely distinct. [It is a condition with the Society of Arts, never to give a reward to any thing for which a patent has been, or is to be, taken out.]

Let us now consider the case of platina. A new process was discovered of rendering it malleable, and the mere circumstance of so large a quantity having been sent into the market, was a positive benefit, of no ordinary magnitude, to many of the arts. The discoverer of this valuable process selected that course for which no reasonable man could blame him; and from some circumstance, or perhaps from accident, he preserved no written record of the manipulations. Had Providence appointed for that disorder, which terminated too fatally, a more rapid career, all the knowledge he had acquired from the long attention he had devoted to the subject, would have been lost to mankind. The hand of a friend recorded the directions of the expiring philosopher, whose anxiety to render useful even his unfinished speculations, proves that the previous omission was most probably accidental.

Under such circumstances it was published to the world in the Transactions of the Royal Society. But what could induce that body to bestow on it their medal? To talk of adding lustre to the name of Wollaston by their medal, is to talk idly. They must have done it then as an example, as a stimulus to urge future inquiries in the career of discovery. But did they wish discoveries to be so endangered?

The discoveries of Professor Mitscherlick, of Berlin, had long been considered, by a few members of the Society, as having strong claims on one of its honorary rewards; but difficulties had arisen, from so few members of the Council having any knowledge of discoveries which had long been familiar to Europe. The Council were just on the point of doing justice to the merits of the Prussian philosopher, when it was suggested that its medal should be given to Dr. Wollaston, and they immediately altered their intention, and thus enabled themselves to reserve their medal to Professor Mitscherlick for another year; at which period, for aught they knew, his discoveries might possess the additional merit of having been made prior to the limit allowed by their regulations. That medal was, in fact, voted at a meeting, at which no one member present was at all conversant with the subjects rewarded. I shall, however, say no more on this subject. They erred from feeling, an error so very rare with them, that it might be pardoned even for its singularity.

I will, however, add one word to those whose censures have been unjustly dealt, to those who have reproached the philosopher for receiving pecuniary advantage from his inventions.

Amongst the many and varied contrivances for the demands of science, or the arts of life, with which we were enriched by the genius of Wollaston, was it too much to allow him to retain, during his fleeting career, one out of the multitude, to furnish that: pecuniary supply, without which, the man will want food for his body, and the philosopher be destitute of tools for his inventions? Had he been, as, from the rank he held in science, he certainly would have been in other kingdoms, rich in the honours his country could bestow, and receiving from her a reward in some measure commensurate with his deserts,—then, indeed, there might have been reason for that reproach; but I am convinced that, in such circumstances, the philosopher would have balanced, with no "niggard" hand, the claims of his country, and would have given to it, unreservedly, the produce of his powerful mind.

SECTION 9. OF THE FAIRCHILD LECTURE.

Mr. Fairchild left by will twenty-five pounds to the Royal Society. This was increased by several subscriptions, and 100L. 3 per cent. South Sea Annuities was purchased, the interest of which was to be devoted annually to pay for a sermon to be preached at St.Leonard's, Shoreditch.

Few members of the Society, perhaps, are aware, either of the bequest or of its annual payment. I shall merely observe, that for five years, from 1800 to 1804, it was regularly given to Mr. Ascough; and that for twenty-six years past, it has been as regularly given to the Rev. Mr. Ellis.

The annual amount is too trifling to stimulate to any extraordinary exertions; yet, small as it is, it might, if properly applied, be productive of much advantage to religion, and of great honour to the Society. For this purpose, it would be desirable that it should be delivered at some church or chapel, more likely to be attended by members of the Royal Society. Notice of it should be given at the place of worship appointed, at least a week previous to its delivery, and at the two preceding weekly meetings of the Royal Society. The name of the gentleman nominated for that year, and the church at which the sermon is to be preached, should be stated.

With this publicity attending it, and by a judicious selection of the first two or three gentlemen appointed to deliver it, it would soon be esteemed an honour to be invited to compose such a lecture, and the Society might always find in its numerous list of members or aspirants, persons well qualified to fulfil a task as beneficial for the promotion of true religion, as it ever must be for the interest of science. I am tempted to believe that such a course would call forth exertions of the most valuable character, as well as give additional circulation to what is already done on that subject.

The geological speculations which have been adduced, perhaps with too much haste by some, as according with the Mosaic history, and by others, as inconsistent with its truth, would, if this subject had been attentively considered, have been allowed to remain until the fullest and freest inquiry had irrevocably fixed their claim to the character of indisputable facts. But, I will not press this subject further on my reader's attention, lest he should think I am myself delivering the lecture. All that I could have said on this point has been so much more ably stated by one whose enlightened view of geological science has taken away some difficulties from its cultivators, and, I hope, removed a stumbling-block from many respectable individuals, that I should only weaken by adding to the argument. [I allude to the critique of Dr. Ure's Geology in the British Review, for July, 1829; an Essay, equally worthy of a philosopher and a Christian.]

SECTION 10. OF THE CROONIAN LECTURE.

The payment [Three pounds.] for this Lecture, like that of the preceding, is small. It was instituted by Dr. Croone, for an annual essay on the subject of Muscular Motion. It is a little to be regretted, that it should have been so restricted; and perhaps its founder, had he foreseen the routine into which it has dwindled, might have endeavoured to preserve it, by affording it a wider range.

By giving it to a variety of individuals, competition might have been created, and many young anatomists have been induced to direct their attention to the favourite inquiry of the founder of the Lecture; but from causes which need not here be traced, this has not been the custom—one individual has monopolized it year after year, and it seems, like the Fairchild Lecture, rather to have been regarded as a pension. There have, however, been some intervals; and we are still under obligations to those who have supported THE SYSTEM, for not appointing Sir Everard Home to read the Croonian Lecture twenty years in SUCCESSION. Had it been otherwise, we might have heard of vested rights.

SECTION 11. OF THE CAUSES OF THE PRESENT STATE OF THE ROYAL SOCIETY.

The best friends of the Royal Society have long admitted, whilst they regretted, its declining fame; and even those who support whatever exists, begin a little to doubt whether it might not possibly be amended.

The great and leading cause of the present state to which the Royal Society is reduced, may be traced to years of misrule to which it has been submitted. In order to understand this, it will be necessary to explain the nature of that misrule, and the means employed in perpetuating it.

It is known, that by the statutes, the body of the Society have the power of electing, annually, their President, Officers, and Council; and it is also well known, that this is a merely nominal power, and that printed lists are prepared and put into the hands of the members on their entering the room, and thus passed into the balloting box. If these lists were, as in other scientific societies, openly discussed in the Council, and then offered by them as recommendations to the Society, little inconvenience would arise; but the fact is, that they are private nominations by the President, usually without notice, to the Council, and all the supporters of the system which I am criticizing, endeavour to uphold the right of this nomination in the President, and prevent or discourage any alteration.

The Society has, for years, been managed by a PARTY, or COTERIE, or by whatever other name may be most fit to designate a combination of persons, united by no expressed compact or written regulations, but who act together from a community of principles. That each individual has invariably supported all the measures of the party, is by no means the case; and whilst instances of opposition amongst them have been very rare, a silent resignation to circumstances has been the most usual mode of meeting measures they disapproved. The great object of this, as of all other parties, has been to maintain itself in power, and to divide, as far as it could, all the good things amongst its members. It has usually consisted of persons of very moderate talent, who have had the prudence, whenever they could, to associate with themselves other members of greater ability, provided these latter would not oppose the system, and would thus lend to it the sanction of their name. The party have always praised each other most highly—have invariably opposed all improvements in the Society, all change in the mode of management; and have maintained, that all those who wished for any alteration were factious; and, when they discovered any symptoms of independence and inquiry breaking out in any member of the Council, they have displaced him as soon as they decently could.

Of the arguments employed by those who support the SYSTEM OF MANAGEMENT by which the Royal Society is governed, I shall give a few samples: refutation is rendered quite unnecessary—juxta-position is alone requisite. If any member, seeing an improper appointment in contemplation, or any abuse in the management of the affairs of the Society continued, raise a voice against it, the ready answer is, Why should you interfere? it may not be quite the thing you approve; but it is no affair of yours.—If, on the other hand, it do relate to himself, the reply is equally ready. It is immediately urged: The question is of a personal nature; you are the last person who ought to bring it forward; you are yourself interested. If any member of the Society, feeling annoyed at the neglect, or hurt by the injuries or insults of the Council, show signs of remonstrance, it is immediately suggested to him that he is irritated, and ought to wait until his feelings subside, and he can judge more coolly on the subject; whilst with becoming candour they admit the ill-treatment, but urge forbearance. If, after an interval, when reflection has had ample time to operate, the offence seems great as at first, or the insult appears unmitigated by any circumstances on which memory can dwell,—if it is then brought forward, the immediate answer is, The affair is out of date—the thing is gone by—it is too late to call in question a transaction so long past. Thus, if a man is interested personally, he is unfit to question an abuse; if he is not, is it probable that he will question it? and if, notwithstanding this, he do so, then he is to be accounted a meddler. If he is insulted, and complain, he is told to wait until he is cool; and when that period arrives, he is then told he is too late. If his remonstrance relates to the alteration of laws which are never referred to, or only known by their repeated breach, he is told that any alteration is useless; it is perfectly well known that they are never adhered to. If it relate to the impolicy of any regulations attaching to an office, he is immediately answered, that that is a personal question, in which it is impossible to interfere—the officer, it seems, is considered to have not merely a vested right to the continuance of every abuse, but an interest in transmitting it unimpaired to his successors.

In the same spirit I have heard errors of calculation or observation defended. If small errors occur, it is said that they are too trifling to be of any importance. If larger errors are pointed out, it is immediately contended that they can deceive nobody, because of their magnitude. Perhaps it might be of some use, if the Council would oblige the world with their SCALE of ERROR, with illustrations from some of the most RECENT and APPROVED works, and would favour the uninformed with the orthodox creed upon all grades, from that which baffles the human faculties to detect, up to that which becomes innocuous from its size.

The offices connected with the Royal Society are few in number, and their emolument small in amount; but the proper disposition of them is, nevertheless, of great importance to the Society, and was so to the science of England.

In the first place, the President, having in effect the absolute nomination of the whole Council, could each year introduce a few gentlemen, whose only qualification to sit on it would be the high opinion they must necessarily entertain of the penetration of him who could discover their scientific merits. He might also place in the list a few nobles or officials, just to gild it. Neither of these classes would put any troublesome questions, and one of them might be employed, from its station in society, to check any that might be proposed by others.

With these ingredients, added to the regular train of the party, and a star or two of science to shed lustre over the whole, a very manageable Council might be formed; and such has been its frequent composition.

The duties of the Secretaries, when well executed, are laborious, although not in this respect equal to those of the same officers who, in several societies, give their gratuitous aid; and their labours are much lightened by the Assistant Secretary and his clerk. The following are their salaries:—

The Senior Secretary .... 105L.
The Junior Secretary, 105L........)
5L. for making Indexto Phil. Trans...) 110L.
The Foreign Secretary........... 20L.

Now it is not customary to change these annually; and as these offices are amongst the "loaves and fishes" they are generally given by the President to some staunch supporters of the system. They have frequently been bestowed, with very little consideration for the interest, or even for the dignity of the Society. To notice only one instance: the late Sir Joseph Banks appointed a gentleman who remained for years in that situation, although he was confessedly ignorant of every subject connected with the pursuits of the Society. I will, however, do justice to his memory, by saying that his respectability was preserved under such circumstances, by the most candid admission of the fact, accompanied by a store of other knowledge unfortunately quite foreign to the pursuits of the Society; and I will add, that I regretted to see him insulted by one President in a situation improperly given to him by a former.

Next in order come the Vice-Presidents, who are appointed by the President; and in this respect the present practice is not inconvenient.

The case, however, is widely different with the office of Treasurer. The President ought not to usurp the power of his appointment, which ought, after serious discussion by the Council, to be made by the Society at large.

Besides the three Secretaries, there is an Assistant Secretary, and recently another has been added, who may perhaps be called a, Sub-assistant Secretary. All these places furnish patronage to the President.

Let us now look at the occasional patronage of the President, arising from offices not belonging to the Society. He is, EX OFFICIO, a Trustee of the British Museum; and it may seem harsh to maintain that he is not a fit person to hold such a situation. It is no theoretical view, but it is the EXPERIENCE of the past which justifies the assertion; and I fear that unless he has the sole responsibility for some specific appointments, and unless his judgment is sharpened by the fear of public discussion, a President of the Royal Society, in the Board-room of the British Museum, is quite as likely as another person to sacrifice his public duty to the influence of power, or to private friendship. With respect to the merits of that Institution, I have no inclination at present to inquire: but when it is considered that there is at this moment attached to it no one whose observations or whose writings have placed him even in the second rank amongst the naturalists of Europe, the President of the Royal Society has given some grounds for the remark made by several members of the Society, that he is a little too much surrounded by the officers of a body who may reasonably be supposed to entertain towards him feelings either of gratitude or expectation. [It will be remembered that the name of Mr. Robert Brown has been but recently attached to the British Museum, and that it is to be attributed to his possessing a life interest in the valuable collection of the late Sir Joseph Banks.]

The late Board of Longitude was another source of patronage, which, although now abolished, it may be useful to hint at.

There were three members to be appointed by the Royal Society: these were honorary, and, as no salary was attached, it might have been expected that this limited number of appointments would have been given in all cases to persons qualified for them. But no: it was convenient to pay compliments; and Lord Colchester, whose talents and knowledge insured him respect as Speaker of the House of Commons, or as a British nobleman, was placed for years in the situation as one of the Commissioners of the Board of Longitude, for which every competent judge knew him to be wholly unfit. What was the return which he made for this indulgence? Little informed respecting the feelings of the Society, and probably misinformed by the party whose influence had placed him there, he saved them in the day of their peril.

When the state of the Society had reached such a point that many of the more scientific members felt that some amendment was absolutely necessary to its respectability, a committee was formed to suggest to the Council such improvements as they might consider it expedient to discuss. [Amongst the names of the persons composing this Committee, which was proposed by Mr. South, were those of Dr. Wollaston and Mr. Herschel.] The Council received their report at the close of the session; and in recording it on the journals, they made an appeal to the Council for the ensuing year to bestow on it "THEIR EARLIEST AND MOST SERIOUS ATTENTION."

Now when the party, to whose government some of these improvements would have been a death-warrant, found that the subject was likely to be taken up in the Council, they were in dismay: but the learned and grateful peer came to their assistance, and aided Mr. Davies Gilbert in getting rid of these improvements completely.

It has been the fashion to maintain that all classes of the Royal Society should be represented in the Council, and consequently that a peer or two should find a place amongst them. Those who are most adverse to this doctrine would perhaps be the most anxious to render this tribute to any one really employing his time, his talents, or his rank in advancing the cause of science. But when a nobleman, unversed in our pursuits, will condescend to use the influence of his station in aiding a President to stifle, WITHOUT DISCUSSION, propositions recommended for consideration by some of the most highly gifted members of the Society,—those who doubt the propriety of the principle may reasonably be pardoned for the disgust they must necessarily entertain for the practical abuse to which it leads.

Of the other three Commissioners, who received each a hundred a-year, although the nomination was, in point of form, in the Admiralty, yet it was well known that the President of the Royal Society did, in fact, always name them. Of these I will only mention one fact. The late Sir Joseph Banks assigned to me as a reason why I need not expect to be appointed, (as he had held out to me at a former period when I had spoken to him on the subject) that I had taken a prominent part in the formation of the ASTRONOMICAL SOCIETY. I am proud of the part I did take in establishing that Society, although an undue share of its honour was assigned to me by the President.

It may, perhaps, be inquired, why I publish this fact at this distance of time? I answer, that I stated it publicly at the Council of the Astronomical Society; —that I always talked of it publicly and openly at the time;—that I purposely communicated it to each succeeding President of the Royal Society; and that, although some may have forgotten the communications I made at the time, there are others who remember them well.

The Secretary of the late Board of Longitude received 300L., and 200L. more, as Superintendent of the Nautical Almanac.

Another situation, in the patronage of which the President is known to have considerable influence, is that of Astronomer Royal; and it is to be observed, that he is kept in the Council as much as possible, notwithstanding the nature of his duties.

Of the three appointments of 100L. a-year each, which have been instituted since the abolition of the Board of Longitude, the President is supposed to have the control, thus making him quite sure of the obedience of his Council.

Besides these sources of patronage, there are other incidental occasions on which Government apply to the Royal Society to recommend proper persons to make particular experiments or observations; and, although I am far from supposing that these are in many instances given to persons the second or third best qualified for them, yet they deserve to be mentioned.

SECTION 12. OF THE PLAN FOR REFORMING THE SOCIETY.

The indiscriminate admission of every candidate became at last so notorious, even beyond the pale of the Society, that some of the members began to perceive the inconveniences to which it led. This feeling, together with a conviction that other improvements were necessary to re-establish the Society in public opinion, induced several of the most active members to wish for some reform in its laws and proceedings; and a Committee was appointed to consider the subject. It was perfectly understood, that the object of this Committee was to inquire,—First, as to the means and propriety of limiting the numbers of this Society; and then, as to other changes which they might think beneficial. The names of the gentlemen composing this Committee were:—

Dr. Wollaston,	Mr. Herschel,
Dr. Young,	Mr. Babbage,
Mr. Davies Gilbert,	Captain Beaufort,
Mr. South,	Captain Kater.

The importance of the various improvements suggested was different in the eyes of different members. The idea of rendering the Society so select as to make it an object of ambition to men of science to be elected into it, was by no means new, as the following extract from the Minutes of the Council will prove:—

"MINUTES OF COUNCIL. August 27, 1674 Present,

Sir W. Petty, Vice-President,
Sir John Lowther,
Sir John Cutler,
Sir Christopher Wren,
Mr. Oldenburgh,
Sir Paul Neile.

"It was considered by this Council, that to make the Society prosper, good experiments must be in the first place provided to make the weekly meetings considerable, and that the expenses for making these experiments must be secured by legal subscriptions for paying the contributors; which done, the Council might then with confidence proceed to the EJECTION OF USELESS FELLOWS."

The reformers of modern times were less energetic in the measures they recommended. Dr. Wollaston and some others thought the limitation of the numbers of the Society to be the most essential point, and 400 was suggested as a proper number to be recommended, in case a limitation should be ultimately resolved upon. I confess, such a limit did not appear to me to bring great advantages, especially when I reflected how long a time must have elapsed before the 714 members of the Society could be reduced by death to that number. And I also thought that as long as those who alone sustained the reputation of the Society by their writings and discoveries should be admitted into it on precisely the same terms, and on the payment of the same sum of money as other gentlemen who contributed only with their purse, it could never be an object of ambition to any man of science to be enrolled on its list.

With this view, and also to assist those who wished for a limitation, I suggested a plan extremely simple in its nature, and which would become effective immediately. I proposed that, in the printed list of the Royal Society, a star should be placed against the name of each Fellow who had contributed two or more papers which had been printed in the Transactions, or that such a list should be printed separately at the end.

At that period there were 109 living members who had contributed papers to the Transactions, and they were thus arranged:

37 Contributors of.. 1 paper
21.......... 2 papers
19.......... 3 ditto
5 .... 4 ditto
3 .... 5 ditto
3 .... 6 ditto
]2.... from 7 to 12 ditto
14... of more than 12 papers.

100 Contributing Fellows of the Royal Society.
589 Papers contributed by them.

Now the immediate effect of printing such a list would be the division of the Society into two classes. Supposing two or more papers necessary for placing a Fellow in the first class, that class would only consist of seventy-two members, which is nearly the same as the number of those of the Institute of France. If only those who had contributed three or more were admitted, then this class would be reduced to fifty-one. In either of these cases it would obviously become a matter of ambition to belong to the first class; and a more minute investigation into the value of each paper would naturally take place before it was admitted into the Transactions. Or it might be established that such papers only should be allowed to count, as the Committee, who reported them as fit to be printed, should also certify. The great objection made to such an arrangement was, that it would be displeasing to the rest of the Society, and that they had a vested right (having entered the Society when no distinction was made in the lists) to have them always continued without one.

Without replying to this shadow of an argument of vested rights, I will only remark that he who maintains this view pays a very ill compliment to the remaining 600 members of the Royal Society; since he does, in truth, maintain that those gentlemen who, from their position, accidentally derive reputation which does not belong to them, are unwilling, when the circumstance is pointed out, to allow the world to assign it to those who have fairly won it; or else that they are incapable of producing any thing worthy of being printed in the Transactions of the Royal Society. Lightly as the conduct of the Society, as a body, has compelled me to think of it, I do not think so ill of the personal character of its members as to believe that if the question were fairly stated to them, many would object to it.

Amongst the alterations which I considered most necessary to the renovation of the Society, was the recommendation, by the expiring Council, of those whom they thought most eligible for that of the ensuing year.

The system which had got into practice was radically bad: it is impossible to have an INDEPENDENT Council if it is named by ONE PERSON. Our statutes were framed with especial regard to securing the fitness of the members elected to serve in the Council; and the President is directed, by those statutes, at the two ordinary meetings previous to the anniversary, to give notice of the elections, and "to declare how much it importeth the good of the Society that such persons may be chosen into the Council as are most likely to attend the meetings and business of the Council, and out of whom may be made the best choice of a President and other officers." This is regularly done; and, in mockery of the wisdom of our ancestors, the President has perhaps in his pocket the list of the future Council he has already fixed upon.

In some other Societies, great advantage is found to arise from the discussion of the proper persons to be recommended to the Society for the Council of the next year. A list is prepared, by the Secretary, of the old Council, and against each name is placed the number of times he has attended the meetings of the Council. Those whose attendance has been least frequent are presumed to be otherwise engaged, unless absence from London, or engagement in some pursuit connected with the Society, are known to have interfered. Those members who have been on the Council the number of years which is usually allowed, added to those who go out by their own wish, and by non-attendance, are, generally, more in number than can be spared; and the question is never, who shall retire?—but, who, out of the rest of the Society, is most likely to work, if placed on the Council?

If any difference of opinion should exist in a society, it is always of great importance to its prosperity to have both opinions represented in the Council. In this age of discussion it is impossible to stifle opinions; and if they are not represented in the Council, there is some chance of their being brought before the general body, or, at last, even before the public. It is certainly an advantage that questions should be put, and even that debates should take place on the days appropriated to the anniversaries of societies. This is the best check to the commencement of irregularities; and a suspicion may reasonably be entertained of those who endeavour to suppress inquiry.

On the other hand, debates respecting the affairs of the Society should never be entered on at the ordinary meetings, as they interrupt its business, and only a partial attendance can be expected. That the conduct of those who have latterly managed the Royal Society has not led to such discussions, is to be attributed more to the forbearance of those who disapprove of the line of conduct they have pursued, than to the discretion of the party in not giving them cause.

The public is the last tribunal; one to which nothing but strong necessity should induce an appeal. There are, however, advantages in it which may, in some cases, render it better than a public discussion at the anniversary. When the cause of complaint is a system rather than any one great grievance, it may be necessary to enter more into detail than a speech will permit; also the printed statement and arguments will probably come under the consideration of a larger number of the members. Another and a considerable benefit is, that there is much less danger of any expression of temper interrupting or injuring the arguments employed.

There were other points suggested, but I shall subjoin the Report of the Committee:—

REPORT OF THE COMMITTEE APPOINTED TO CONSIDER THE BEST MEANS OF LIMITING THE MEMBERS ADMITTED INTO THE ROYAL SOCIETY, AS WELL AS TO MAKE SUCH SUGGESTIONS ON THAT SUBJECT AS MAY SEEM TO THEM CONDUCIVE TO THE WELFARE OF THE SOCIETY.

Your Committee having maturely considered the resolution of the Council under which they have been appointed; and having satisfied themselves that the progressive increase of the Society has been in a much higher ratio than the progressive increase of population, or the general growth of knowledge, or the extension of those sciences which it has been the great object of the Society to promote, they have agreed to the following Report:—

Your Committee assume as indisputable propositions, that the utility of the Society is in direct proportion to its respectability. That its respectability can only be secured by its comprising men of high philosophical eminence; and that the obvious means of associating persons of this eminence will be the public conviction, that to belong to the Society is an honour. Your Committee, therefore, think themselves fully borne out in the conclusion, that it would be expedient to limit the Society to such a number as should be a fair representation of the talent of the country; the consequence of which will be, that every vacancy would become an object of competition among persons of acknowledged merit.

From the returns which have been laid on your table, of the Fellows who have contributed papers, and from the best estimate they can make of the persons without doors who are engaged in the active pursuit of science, your Committee feel justified in recommending that those limits should be fixed at four hundred, exclusive of foreign members, and of such royal personages as it may be thought proper to admit.

As many years must elapse before the present number of seven hundred and fourteen can be reduced to those limits by the course of nature, and as it would be prejudicial to the interests of the Society and of science, that no fresh accessions should take place during that long period, your Committee would further recommend, that till that event takes place, four new members should be annually admitted.

With respect to the manner of admission, your Committee are of opinion, that there are several inconveniences in the present mode of proceeding to a single ballot upon each certificate, according to its seniority. If the above limitation should be adopted, it may be presumed, that for every vacancy there will be many candidates; from amongst them, it must be the general wish to select the most distinguished individuals; but to accomplish this, if the present system were to be continued, it would be necessary to reject all those candidates whose certificates were of earlier date than theirs; a process not only extremely irritating, but probably ineffectual from the want of unanimity. Your

Committee, therefore, most earnestly recommend, that one general election should take place every year towards the end of the session, and that this should be conducted on the same principles as the present annual election of the Council and officers; VIZ. by having lists printed of all the candidates (whose certificates had been suspended for the usual time,) in which lists each Fellow would mark the requisite number of persons.

As the charter, however, requires the concurrence of two-thirds of the Fellows present, your Committee suggest, that after the choice has been determined by the plurality of votes by ballot in the above manner, the successful candidates should be again submitted to a general vote, in accordance with the enactments of the said charter.

In concluding this part of the subject, your Committee beg leave to remark, that by the method now proposed, the invidious act of blackballing would cease, and with it all feelings of resentment and mortification; as the result of such an open competition could only be construed by the public into a fair preference of the superior claims of the successful few, and not into a direct and disgraceful rejection of the others.

Your Committee are fully aware, that such a reduction in the usual admissions would materially affect the pecuniary resources of the Society; but they are at the same time convinced, that by a vigorous economy its present income might be rendered adequate to all its real wants, and the aggregate expenditure might be considerably diminished by many small but wholesome retrenchments.

It appears, from the accounts of last year, that although 1200L. was received for compositions, in addition to the standing income, and usual contributions, &c., and although no money was invested, yet there was a balance only of a few pounds at the end of the year. It further appears, that 500L. was paid for the paper, 370L. for engravings, and nearly 340L. for printing; and from those alarming facts, your Committee submit to your consideration, whether the expenditure might not be beneficially controlled by a standing Committee of Finance.

In obedience to the latter part of your resolution, your Committee now proceed to offer some further suggestions for your consideration. They conceive that it would afford a beneficial stimulus to individual exertion, if the Fellows who have received the medals of the Society, and those who have repeatedly enriched its Transactions, were distinguished by being collected into a separate and honourable list. It would also be found, perhaps, not less a future incentive than an act of retrospective justice, if the names of all those illustrious Fellows who have formerly obtained the medals, as well as of all those individuals who have been large benefactors to the Society, were recorded at the end of the list. It would be a satisfactory addition likewise to the annual list, if all those Fellows who have died, or had been admitted within the preceding year, were regularly noticed. And your Committee think, that these lists should always form part of the Transactions, and be stitched up with the last part of the volume.

It requires no argument to demonstrate that the well-being of the Society mainly depends on the activity and integrity of its Council; and as their selection is unquestionably of paramount importance, your Committee hope that our excellent President will not consider it any impeachment of his impartiality, or any doubt of his zeal, if they venture to suggest, that the usual recommendation to the Society of proper members for the future Council should henceforth be considered as a fit subject for the diligent and anxious deliberation of the expiring Council.

There is another point of great moment to the character of the Society, and to the dignified station it occupies among the learned associations of Europe; for its character abroad can only be appreciated by the nature and value of its Transactions. Your Committee allude to the important task of deciding on what papers should be published; and they are of opinion that it would be a material improvement on the present mode, if each paper were referred to a separate Committee, who should have sufficient time given them to examine it carefully, who should be empowered to communicate on any doubtful parts with the author; and who should report, not only their opinion, but the grounds on which that opinion is formed, for the ultimate decision of the Council.

If it should be thought fit to adopt the suggestions which your Committee have now had the honour of proposing, they beg leave to move, that another Committee be appointed, with directions to frame or to alter the necessary statutes, so that they may be in strict accordance with the charters.

In concluding the Report, your Committee do not wish to disguise the magnitude of some of the measures they have thought it their duty to propose; on the contrary, they would not only urge the fullest discussion of their expediency; but further, that if you should even be unanimously disposed to confirm them, your Committee would recommend, that the several statutes, when they have been drawn up or modified, should be only entered on your minutes, and not finally enacted. All innovations in the constitution, or even the habits of the Royal Society, should be scrutinized with the most jealous circumspection. It is enough for the present Council to have traced the plan; let the Council of the ensuing sessions share the credit of carrying that plan into effect.

This Report was presented to the Council very late in the session of 1827, and on the 25th of June there occurs the following entry on the council-book: —

"The Report of the Committee for considering the best means of limiting the number of members, and such other suggestions as they may think conducive to the good of the Society, was received and read, and ordered to be entered on the minutes; and the Council, regarding the importance of the subject, and its bearings on the essential interests of the Society, in conformity with the concluding paragraph, and considering also the advanced stage of the session, recommend it to the most serious and early consideration of the Council for the ensuing year."

Those who advocated these alterations, were in no hurry for their hasty adoption; they were aware of their magnitude, and anxious for the fullest investigation before one of them should be tried.

Unfortunately, the concluding recommendation of the Committee did not coincide with the views of Mr. Gilbert, whom the party had determined to make their new President. That gentleman made such arrangements for the Council of the succeeding year, that when the question respecting the consideration of the Report of that Committee was brought forward, it was thrown aside in the manner I have stated. Thus a report, sanctioned by the names of such a committee, and recommended by one Council to "THE MOST SERIOUS and EARLY consideration of the Council for the ensuing year," was by that very Council rejected, without even the ceremony of discussing its merits. Was every individual recommendation it contained, not merely unfit to be adopted, but so totally deficient in plausibility as to be utterly unworthy of discussion? Or did the President and his officers feel, that their power rested on an insecure foundation, and that they did not possess the confidence of the working members of the Society?

CHAPTER V. OF OBSERVATIONS.

There are several reflections connected with the art of making observations and experiments, which may be conveniently arranged in this chapter.

SECTION 1. OF MINUTE PRECISION.

No person will deny that the highest degree of attainable accuracy is an object to be desired, and it is generally found that the last advances towards precision require a greater devotion of time, labour, and expense, than those which precede them. The first steps in the path of discovery, and the first approximate measures, are those which add most to the existing knowledge of mankind.

The extreme accuracy required in some of our modern inquiries has, in some respects, had an unfortunate influence, by favouring the opinion, that no experiments are valuable, unless the measures are most minute, and the accordance amongst them most perfect. It may, perhaps, be of some use to show, that even with large instruments, and most practised observers, this is but rarely the case. The following extract is taken from a representation made by the present Astronomer-Royal, to the Council of the Royal Society, on the advantages to be derived from the employment of two mural circles:—

"That by observing, with two instruments, the same objects at the same time, and in the same manner, we should be able to estimate how much of that OCCASIONAL DISCORDANCE FROM THE MEAN, which attends EVEN THE MOST CAREFUL OBSERVATIONS, ought to be attributed to irregularity of refraction, and how much to THE IMPERFECTIONS OF INSTRUMENTS."

In confirmation of this may be adduced the opinion of the late M. Delambre, which is the more important, from the statement it contains relative to the necessity of publishing all the observations which have been made.

"Mais quelque soit le parti que l'on prefere, il me semble qu'on doit tout publier. Ces irregularites memes sont des faits qu'il importe de connoitre. LES SOINS LES PLUS ATTENTIFS N'EN SAUROIENT PRESERVER LES OBSERVATEURS LES PLUS EXERCES, et celui qui ne produiroit que des angles toujours parfaitment d'accord auroit ete singulierement bien servi par les circonstances ou ne seroit pas bien sincere."—BASE DU SYSTEME METRIQUE, Discours Preliminaire, p. 158.

This desire for extreme accuracy has called away the attention of experimenters from points of far greater importance, and it seems to have been too much overlooked in the present day, that genius marks its tract, not by the observation of quantities inappreciable to any but the acutest senses, but by placing Nature in such circumstances, that she is forced to record her minutest variations on so magnified a scale, that an observer, possessing ordinary faculties, shall find them legibly written. He who can see portions of matter beyond the ken of the rest of his species, confers an obligation on them, by recording what he sees; but their knowledge depends both on his testimony and on his judgment. He who contrives a method of rendering such atoms visible to ordinary observers, communicates to mankind an instrument of discovery, and stamps his own observations with a character, alike independent of testimony or of judgment.

SECTION 2. ON THE ART OF OBSERVING.

The remarks in this section are not proposed for the assistance of those who are already observers, but are intended to show to persons not familiar with the subject, that in observations demanding no unrivalled accuracy, the principles of common sense may be safely trusted, and that any gentleman of liberal education may, by perseverance and attention, ascertain the limits within which he may trust both his instrument and himself.

If the instrument is a divided one, the first thing is to learn to read the verniers. If the divisions are so fine that the coincidence is frequently doubtful, the best plan will be for the learner to get some acquaintance who is skilled in the use of instruments, and having set the instrument at hazard, to write down the readings of the verniers, and then request his friend to do the same; whenever there is any difference, he should carefully examine the doubtful one, and ask his friend to point out the minute peculiarities on which he founds his decision. This should be repeated frequently; and after some practice, he should note how many times in a hundred his reading differs from his friend's, and also how many divisions they usually differ.

The next point is, to ascertain the precision with which the learner can bisect an object with the wires of the telescope. This can be done without assistance. It is not necessary even to adjust the instrument, but merely to point it to a distant object. When it bisects any remarkable point, read off the verniers, and write down the result; then displace the telescope a little, and adjust it again. A series of such observations will show the confidence which is due to the observer's eye in bisecting an object, and also in reading the verniers; and as the first direction gave him some measure of the latter, he may, in a great measure, appreciate his skill in the former. He should also, when he finds a deviation in the reading, return to the telescope, and satisfy himself if he has made the bisection as complete as he can. In general, the student should practise each adjustment separately, and write down the results wherever he can measure its deviations.

Having thus practised the adjustments, the next step is to make an observation; but in order to try both himself and the instrument, let him take the altitude of some fixed object, a terrestrial one, and having registered the result, let him derange the adjustment, and repeat the process fifty or a hundred times. This will not merely afford him excellent practice, but enable him to judge of his own skill.

The first step in the use of every instrument, is to find the limits within which its employer can measure the SAME OBJECT UNDER THE SAME CIRCUMSTANCES. It is only from a knowledge of this, that he can have confidence in his measures of the SAME OBJECT UNDER DIFFERENT CIRCUMSTANCES, and after that, of DIFFERENT OBJECTS UNDER DIFFERENT CIRCUMSTANCES.

These principles are applicable to almost all instruments. If a person is desirous of ascertaining heights by a mountain barometer, let him begin by adjusting the instrument in his own study; and having made the upper contact, let him write down the reading of the vernier, and then let him derange the UPPER adjustment ONLY, re-adjust, and repeat the reading. When he is satisfied about the limits within which he can make that adjustment, let him do the same repeatedly with the lower; but let him not, until he knows his own errors in reading and adjusting, pronounce upon those of the instrument. In the case of a barometer, he must also be assured, that the temperature of the mercury does not change during the interval.

A friend once brought to me a beautifully constructed piece of mechanism, for marking minute portions of time; the three-hundredth parts of a second were indicated by it. It was a kind of watch, with a pin for stopping one of the hands. I proposed that we should each endeavour to stop it twenty times in succession, at the same point. We were both equally unpractised, and our first endeavours showed that we could not be confident of the twentieth part of a second. In fact, both the time occupied in causing the extremities of the fingers to obey the volition, as well as the time employed in compressing the flesh before the fingers acted on the stop, appeared to influence the accuracy of our observations. From some few experiments I made, I thought I perceived that the rapidity of the transmission of the effects of the will, depended on the state of fatigue or health of the body. If any one were to make experiments on this subject, it might be interesting, to compare the rapidity of the transmission of volition in different persons, with the time occupied in obliterating an impression made on one of the senses of the same persons. For example, by having a mechanism to make a piece of ignited charcoal revolve with different degrees of velocity, some persons will perceive a continuous circle of light before others, whose retina does not retain so long impressions that are made upon it.

SECTION 3. ON THE FRAUDS OF OBSERVERS.

Scientific inquiries are more exposed than most others to the inroads of pretenders; and I feel that I shall deserve the thanks of all who really value truth, by stating some of the methods of deceiving practised by unworthy claimants for its honours, whilst the mere circumstance of their arts being known may deter future offenders.

There are several species of impositions that have been practised in science, which are but little known, except to the initiated, and which it may perhaps be possible to render quite intelligible to ordinary understandings. These may be classed under the heads of hoaxing, forging, trimming, and cooking.

OF HOAXING. This, perhaps, will be better explained by an example. In the year 1788, M. Gioeni, a knight of Malta, published at Naples an account of a new family of Testacea, of which he described, with great minuteness, one species, the specific name of which has been taken from its habitat, and the generic he took from his own family, calling it Gioenia Sicula. It consisted of two rounded triangular valves, united by the body of the animal to a smaller valve in front. He gave figures of the

161

animal, and of its parts; described its structure, its mode of advancing along the sand, the figure of the tract it left, and estimated the velocity of its course at about two-thirds of an inch per minute. He then described the structure of the shell, which he treated with nitric acid, and found it approach nearer to the nature of bone than any other shell.

The editors of the ENCYCLOPEDIE METHODIQUE, have copied this description, and have given figures of the Gioenia Sicula. The fact, however, is, that no such animal exists, but that the knight of Malta, finding on the Sicilian shores the three internal bones of one of the species of Bulla, of which some are found on the south-western coast of England, [Bulla lignaria] described and figured these bones most accurately, and drew the whole of the rest of the description from the stores of his own imagination.

Such frauds are far from justifiable; the only excuse which has been made for them is, when they have been practised on scientific academies which had reached the period of dotage. It should however be remembered, that the productions of nature are so various, that mere strangeness is very far from sufficient to render doubtful the existence of any creature for which there is evidence; [The number of vertebrae in the neck of the plesiosaurus is a strange but ascertained fact] and that, unless the memoir itself involves principles so contradictory, as to outweigh the evidence of a single witness, [The kind of contradiction which is here alluded to, is that which arises from well ascertained final causes; for instance, the ruminating stomach of the hoofed animals, is in no case combined with the claw-shaped form of the extremities, frequent in many of the carniverous animals, and necessary to some of them for the purpose of seizing their prey] it can only be regarded as a deception, without the accompaniment of wit.

FORGING differs from hoaxing, inasmuch as in the latter the deceit is intended to last for a time, and then be discovered, to the ridicule of those who have credited it; whereas the forger is one who, wishing to acquire a reputation for science, records observations which he has never made. This is sometimes accomplished in astronomical observations by calculating the time and circumstances of the phenomenon from tables. The observations of the second comet of 1784, which was only seen by the Chevalier D'Angos, were long suspected to be a forgery, and were at length proved to be so by the calculations and reasonings of Encke. The pretended observations did not accord amongst each other in giving any possible orbit. But M. Encke detected an orbit, belonging to some of the observations, from which he found that all the rest might be almost precisely deduced, provided a mistake of a unity in the index of the logarithm of the radius vector were supposed to have been made in all the rest of the calculations. ZACH. CORR. ASTRON. Tom. IV. p. 456.

Fortunately instances of the occurrence of forging are rare.

TRIMMING consists in clipping off little bits here and there from those observations which differ most in excess from the mean, and in sticking them on to those which are too small; a species of "equitable adjustment," as a radical would term it, which cannot be admitted in science.

This fraud is not perhaps so injurious (except to the character of the trimmer) as cooking, which the next paragraph will teach, The reason of this is, that the AVERAGE given by the observations of the trimmer is the same, whether they are trimmed or untrimmed. His object is to gain a reputation for extreme accuracy in making observations; but from respect for truth, or from a prudent foresight, he does not distort the position of the fact he gets from nature, and it is usually difficult to detect him. He has more sense or less adventure than the Cook.

OF COOKING. This is an art of various forms, the object of which is to give to ordinary observations the appearance and character of those of the highest degree of accuracy.

One of its numerous processes is to make multitudes of observations, and out of these to select those only which agree, or very nearly agree. If a hundred observations are made, the cook must be very unlucky if he cannot pick out fifteen or twenty which will do for serving up.

Another approved receipt, when the observations to be used will not come within the limit of accuracy, which it has been resolved they shall possess, is to calculate them by two different formulae. The difference in the constants employed in those formulae has sometimes a most happy effect in promoting unanimity amongst discordant measures. If still greater accuracy is required, three or more formulae can be used.

It must be admitted that this receipt is in some instances rather hazardous: but in cases where the positions of stars, as given in different catalogues, occur, or different tables of specific gravities, specific heats, &c. &c., it may safely be employed. As no catalogue contains all stars, the computer must have recourse to several; and if he is obliged to use his judgment in the selection, it would be cruel to deny him any little advantage which might result from it. It may, however, be necessary to guard against one mistake into which persons might fall.

If an observer calculate particular stars from a catalogue which makes them accord precisely with the rest of his results, whereas, had they been computed from other catalogues the difference would have been considerable, it is very unfair to accuse him of COOKING; for—those catalogues may have been notoriously inaccurate; or—they may have been superseded by others more recent, or made with better instruments; or—the observer may have been totally ignorant of their existence.

It sometimes happens that the constant quantities in formulae given by the highest authorities, although they differ amongst themselves, yet they will not suit the materials. This is precisely the point in which the skill of the artist is shown; and an accomplished cook will carry himself triumphantly through it, provided happily some mean value of such constants will fit his observations. He will discuss the relative merits of formulae he has just knowledge enough to use; and, with admirable candour assigning their proper share of applause to Bessel, to Gauss, and to Laplace, he will take THAT mean value of the constant used by three such philosophers, which will make his own observations accord to a miracle.

There are some few reflections which I would venture to suggest to those who cook, although they may perhaps not receive the attention which, in my opinion, they deserve, from not coming from the pen of an adept.

In the first place, it must require much time to try different formulae. In the next place it may happen that, in the progress of human knowledge, more correct formula: may be discovered, and constants may be determined with far greater precision. Or it may be found that some physical circumstance influences the results, (although unsuspected at the time) the measure of which circumstance may perhaps be recovered from other contemporary registers of facts. [Imagine, by way of example, the state of the barometer or thermometer.] Or if the selection of observations has been made with the view of its agreeing precisely with the latest determination, there is some little danger that the average of the whole may differ from that of the chosen ones, owing to some law of nature, dependent on the interval between the two sets, which law some future philosopher may discover, and thus the very best observations may have been thrown aside.

In all these, and in numerous other cases, it would most probably happen that the cook would procure a temporary reputation for unrivalled accuracy at the expense of his permanent fame. It might also have the effect of rendering even all his crude observations of no value; for that part of the scientific world whose opinion is of most weight, is generally so unreasonable, as to neglect altogether the observations of those in whom they have, on any occasion, discovered traces of the artist. In fact, the character of an observer, as of a woman, if doubted is destroyed.

The manner in which facts apparently lost are restored to light, even after considerable intervals of time, is sometimes very unexpected, and a few examples may not be without their use. The thermometers employed by the philosophers who composed the Academia Del Cimento, have been lost; and as they did not use the two fixed points of freezing and boiling water, the results of a great mass of observations have remained useless from our ignorance of the value of a degree on their instrument. M. Libri, of Florence, proposed to regain this knowledge by comparing their registers of the temperature of the human body and of that of some warm springs in Tuscany, which have preserved their heat uniform during a century, as well as of other things similarly circumstanced.

Another illustration was pointed out to me by M. Gazzeri, the Professor of Chemistry at Florence. A few years ago an important suit in one of the legal courts of Tuscany depended on ascertaining whether a certain word had been erased by some chemical process from a deed then before the court. The party who insisted that an erasure had been made, availed themselves of the knowledge of M. Gazzeri, who, concluding that those who committed the fraud would be satisfied by the disappearance of the colouring matter of the ink, suspected (either from some colourless matter remaining in the letters, or perhaps from the agency of the solvent having weakened the fabric of the paper itself beneath the supposed letters) that the effect of the slow application of heat would be to render some difference of texture or of applied substance evident, by some variety in the shade of colour which heat in such circumstances might be expected to produce. Permission having been given to try the experiment, on the application of heat the important word reappeared, to the great satisfaction of the court.

CHAPTER VI. SUGGESTIONS FOR THE ADVANCEMENT OF SCIENCE IN ENGLAND.

SECTION 1. OF THE NECESSITY THAT MEMBERS OF THE ROYAL SOCIETY SHOULD

EXPRESS THEIR OPINIONS.

One of the causes which has contributed to the success of the PARTY, is to be found in the great reluctance with which many of those whose names added lustre to the Society expressed their opinions, and the little firmness with which they maintained their objections. How many times have those whose activity was additionally stimulated by their interest, proposed measures which a few words might have checked; whilst the names of those whose culpable silence thus permitted the project to be matured, were immediately afterwards cited by their grateful coadjutors, as having sanctioned that which in their hearts they knew to be a job.

Even in the few cases which have passed the limits of such forbearance, when the subject has been debated in the Council, more than one, more than two instances are known, where subsequent circumstances have occurred, which proved, with the most irresistible moral evidence, that members have spoken on one side of the question, and have voted on the contrary.

This reluctance to oppose that which is disapproved, has been too extensively and too fatally prevalent for the interests of the Royal Society. It may partly be attributed to that reserved and retiring disposition, which frequently marks the man of real knowledge, as strongly as an officious interference and flippant manner do the charlatan, or the trader in science. Some portion of it is due to that improper deference which was long paid to every dictum of the President, and much of it to that natural indisposition to take trouble on any point in which a man's own interest is not immediately concerned. It is to be hoped, for the credit of that learned body, that no anticipation of the next feast of St. Andrew ever influenced the taciturnity of their disposition. [It may be necessary to inform those who are not members of the Royal Society, that this is the day on which those Fellows who choose, meet at Somerset House, to register the names of the Council and Officers the President has been pleased to appoint for the ensuing year; and who afterwards dine together, for the purpose of praising each other over wine, which, until within these few years, was PAID for out of the FUNDS of the Society. This abuse was attacked by an enterprising reformer, and of course defended by the coterie. It was, however, given up as too bad. The public may form some idea of the feeling which prevails in the Council, when they are informed that this practice was defended by one of the officers of the Society, on the ground that, if abolished, THE ASSISTANT SECRETARY WOULD LOSE HIS PERCENTAGE ON THE TAVERN BILLS.]

SECTION 2. OF BIENNIAL PRESIDENTS.

The days in which the Royal Society can have much influence in science seem long past; nor does it appear a matter of great importance who conduct its mismanaged affairs. Perpetual Presidents have been tried until the Society has become disgusted with dictators. If any reform should be attempted, it might perhaps be deserving consideration whether the practice of several of the younger institutions might not be worthy imitation, and the office of President be continued only during two sessions. There may be some inconveniences attending this arrangement; but the advantages are conspicuous, both in the Astronomical and Geological Societies. Each President is ambitious of rendering the period of his reign remarkable for some improvement in the Society over which he presides; and the sacrifice of time which is made by the officers of those Societies, would become impossible if it were required to be continued for a much longer period. Another circumstance of considerable importance is, that the personal character of the President is less impressed on the Society; and, supposing any injudicious alterations to be made, it is much less difficult to correct them.

SECTION 3. OF THE INFLUENCE OF THE COLLEGES OF PHYSICIANS AND SURGEONS

IN THE ROYAL SOCIETY.
The honour of belonging to the Royal Society

is much sought after by medical men, as contributing to the success of their professional efforts, and two consequences result from it. In the first place, the pages of the Transactions of the Royal Society occasionally contain medical papers of very moderate merit; and, in the second, the preponderance of the medical interest introduces into the Society some of the jealousies of that profession. On the other hand, medicine is intimately connected with many sciences, and its professors are usually too much occupied in their practice to exert themselves, except upon great occasions.

SECTION 4. OF THE INFLUENCE OF THE ROYAL INSTITUTION ON THE ROYAL SOCIETY.

The Royal Institution was founded for the cultivation of the more popular and elementary branches of scientific knowledge, and has risen, partly from the splendid discoveries of Davy, and partly from the decline of the Royal Society, to a more prominent station than it would otherwise have occupied in the science of England. Its general effects in diffusing knowledge among the more educated classes of the metropolis, have been, and continue to be, valuable. Its influence, however, in the government of the Royal Society, is by no means attended with similar advantages, and has justly been viewed with considerable jealousy by many of the Fellows of that body. It may be stated, without disparagement to the Royal Institution, that the scientific qualifications necessary for its officers, however respectable, are not quite of that high order which ought to be required for those of the Royal Society, if the latter body were in a state of vigour.

The Royal Institution interest has always been sufficient to appoint one of the Secretaries of the Royal Society; and at the present moment they have appointed two. In a short time, unless some effectual check is put to this, we shall find them nominating the President and the rest of the officers. It is certainly not consistent with the dignity of the Royal Society thus to allow its offices to be given away as the rewards of services rendered to other institutions. The only effectual way to put a stop to this increasing interest would be, to declare that no manager or officer of the Royal Institution should ever, at the same time, hold office in the Royal Society.

The use the Members of the Royal Institution endeavour to make of their power in the Council of the Royal Society, is exemplified in the minutes of the Council of March 11, 1830, which may be consulted with advantage by those who doubt.

SECTION 5. OF THE TRANSACTIONS OF THE ROYAL SOCIETY.

The Transactions of the Royal Society, unlike those of most foreign academies, contain nothing relating to the history of the Society. The volumes contain merely those papers communicated to the Society in the preceding year which the Council have selected for printing, a meteorological register, and a notice of the award of the annual medals, without any list of the Council and officers of the Society, by whom that selection and that award have been made.

Before I proceed to criticise this state of things, I will mention one point on which I am glad to be able to bestow on the Royal Society the highest praise. I refer to the extreme regularity with which the volumes of the Transactions are published. The appearance of the half-volumes at intervals of six months, insures for any communication almost immediate publicity; whilst the shortness of the time between its reception and publication, is a guarantee to the public that the whole of the paper was really communicated at the time it bears date. To this may also be added, the rarity of any alterations made previously to the printing, a circumstance which ought to be imitated, as well as admired, by other societies. There may, indeed, be some, perhaps the

Geological, in which the task is more difficult, from the nature of the subject. The sooner, however, all societies can reduce themselves to this rule, of rarely allowing any thing but a few verbal corrections to papers that are placed in their hands, the better it will be for their own reputation, and for the interests of science.

It has been, and continues to be, a subject of deep regret, that the first scientific academy in Europe, the Institute of France, should be thus negligent in the regularity of its publications; and it is the more to be regretted, that it should be years in arrear, from the circumstance, that the memoirs admitted into their collection are usually of the highest merit. I know some of their most active members have wished it were otherwise; I would urge them to put a stop to a practice, which, whilst it has no advantages to recommend it, is unjust to those who contribute, and is only calculated to produce conflicting claims, equally injurious to science, and to the reputation of that body, whose negligence may have given rise to them. [Mr. Herschel, speaking of a paper of Fresnel's, observes —"This memoir was read to the Institute, 7th of October, 1816; a supplement was received, 19th of January, 1818; M. Arago's report on it was read, 4th of June, 1821: and while every optical philosopher in Europe has been impatiently expecting its appearance for seven years, it lies as yet unpublished, and is only known to us by meagre notices in a periodical journal." MR HERSCHEL'S TREATISE ON LIGHT, p. 533.— ENCYCLOPAEDIA METROPOLITANA.]

One of the inconveniences arising from having no historical portion in the volumes of the Royal Society is, that not only the public, but our own members are almost entirely ignorant of all its affairs. With a means of giving considerable publicity (by the circulation of above 800 copies of the Transactions) to whatever we wish to have made known to our members or to the world, will it be credited, that no notice was taken in our volume for 1826, of the foundation of two Royal medals, nor of the conditions under which they were to be distributed. [That the Council refrained from having their first award of those medals thus communicated, is rather creditable to them, and proves that they had a becoming feeling respecting their former errors.] That in 1828, when a new fund, called the donation fund, was established, and through the liberality of Dr. Wollaston and Mr. Davies Gilbert, it was endowed by them with the respective sums of 2,000L. and 1,000L. 3 per cents; no notice of such fact appears in our Transactions for 1829. Other gentlemen have contributed; and if it is desirable to possess such a fund, it is surely of importance to inform the non-attending, which is by far the largest part of the Society, that it exists; and that we are grateful to those by whom it has been founded and augmented. Neither did the Philosophical Transactions inform our absent members, that they could purchase the President's Discourses at the trade-price.

The list of the Officers, Council, and Members of the Royal Society is printed annually; yet, who ever saw it bound up with the Philosophical

Transactions, to which it is intended to be attached? I never met with a single copy of that work so completed, not even the one in our own library. It is extremely desirable that the Society should know the names of their Council; and whilst it would in some measure contribute to prevent the President from placing incompetent persons upon it, it would also afford some check, although perhaps but a slight one, on the distribution of the medals. When I have urged the expediency of the practice, I have been answered by excuses, that the list could not be made up in time for the volume. If this is true of the first part, they might appear with the second; and even if this were impracticable, the plan of prefixing them to the volume of the succeeding year, would be preferable to that of omitting them altogether. The true reason, however, appeared at last. It was objected to the plan, that by the present arrangement, the porter of the Royal Society took round the list to those members resident in London, and got from some of them a remuneration, in the shape of a Christmas-box; and this would be lost, if the time of printing were changed. [During the printing of this chapter, a friend, on whom I had called, complained that the porter of the Royal Society had demanded half-a-crown for leaving the list.] Such are the paltry interests to which those of the Royal Society are made to bow.

Another point on which information ought to be given in each volume, is the conditions on which the distribution of the Society's medals are made. It is true that these are, or ought to be, printed with the Statutes of the Society; but that volume is only in the hands of members, and it is for the credit of the medals themselves, that the laws which regulate their award should be widely known, in order that persons, not members of the Society, might enter into competition for them.

Information relative to the admissions and deaths amongst the Society would also be interesting; a list of the names of those whom the Society had lost, and of those members who had been added to its ranks each year, would find a proper place in the historical pages which ought to be given with each volume of our Transactions.

The want of a distinction between the working members of the Society, and those who merely honour it with their patronage, renders many arrangements, which would be advantageous to science, in some cases, injudicious, and in other instances, almost impossible.

Collections of Observations which are from time to time given to the Society, may be of such a nature, that but few of the members are interested in them. In such cases, the expense of printing above 800 copies may reasonably induce the Council to decline printing them altogether; whereas, if they had any means of discrimination for distributing them, they might be quite willing to incur the expense of printing 250. Other cases may occur, in which great advantage would accrue, if the principle were once admitted. Government, the Universities, public bodies, and even individuals might, in some cases, be disposed to present to the Royal Society a limited number of copies of their works, if they knew that they were likely to be placed in the hands of persons who would use them. Fifty or a hundred additional copies might, in some cases, not be objected to on the ground of expense, when seven or eight hundred would be quite out of the question.

Let us suppose twenty copies of a description of some new chemical process to be placed at the disposal of the Royal Society by any public body; it will not surely be contended that they ought all to remain on the Society's shelves. Yet, with our present rules, that would be the case. If, however, the list of the Members of the Society were read over to the Council, and the names of those gentlemen known to be conversant with chemical science were written down; then, if nineteen copies of the work were given to those nineteen persons on this list, who had contributed most to the Transactions of the Society, they would in all probability be placed in the fittest hands.

Complete sets of the Philosophical Transactions have now become extremely bulky; it might be well worth our consideration, whether the knowledge of the many valuable papers they contain would not be much spread, by publishing the abstracts of them which have been read at the ordinary meetings of the Society. Perhaps two or three volumes octavo, would contain all that has been done in this way during the last century.

Another circumstance, which would contribute much to the order of the proceedings of the Council, would be to have a distinct list made out of all the statutes and orders of the Council relating to each particular subject.

Thus the President, by having at one view

before him all that had ever been decreed on the question under consideration, would be much better able to prevent inconsistent resolutions, and to save the time of the Council from being wasted by unnecessary discussions.

SECTION 6. ORDER OF MERIT.

Amongst the various proposals for encouraging science, the institution of an order of merit has been suggested. It is somewhat singular, that whilst in most of the other kingdoms of Europe, such orders exist for the purpose of rewarding, by honorary distinctions, the improvers of the arts of life, or successful discoverers in science, nothing of the kind has been established in England. [At the great meeting of the philosophers at Berlin, in 1828, of which an account is given in the Appendix; the respect in which Berzelius, Oersted, Gauss, and Humboldt were held in their respective countries was apparent in the orders bestowed on them by the Sovereigns of Sweden, of Denmark, of Hanover, and of Prussia; and there were present many other philosophers, whose decorations sufficiently attested the respect in which science was held in the countries from which they came.]

Our orders of knighthood are favourable only

to military distinction. It has been urged, as an argument for such institutions, that they are a cheap mode of rewarding science, whilst, on the other hand, it has been objected, that they would diminish the value of such honorary distinctions by making them common. The latter objection is of little weight, because the numbers who pursue science are few, and, probably, will long continue so. It would also be easily avoided, by restricting the number of the order or of the class, if it were to form a peculiar class of another order. Another objection, however, appears to me to possess far greater weight; and, however strong the disposition of the Government might be (if such an order existed) to fill it properly, I do not believe that, in the present state of public opinion respecting science, it could be done, and, in all probability, it would be filled up through the channels of patronage, and by mere jobbers in science.

Another proposal, of a similar kind, has also been talked of, one which it may appear almost ridiculous to suggest in England, but which would be considered so in no other country. It is, to ennoble some of the greatest scientific benefactors of their country. Not to mention political causes, the ranks of the nobility are constantly recruited from the army, the navy, and the bar; why should not the family of that man, whose name is imperishably connected with the steam-engine, be enrolled amongst the nobility of his country? In utility and profit, not merely to that country, but to the human race, his deeds may proudly claim comparison even with the most splendid of those achieved by classes so rich in glorious recollections. An objection, in most cases fatal to such a course, arises from the impolicy of conferring a title, unless a considerable fortune exists to support it; a circumstance very rarely occurring to the philosopher. It might in some measure be removed, by creating such titles only for life. But here, again, until there existed some knowledge of science amongst the higher classes, and a sound state of public opinion relative to science, the execution of the plan could only be injurious.

SECTION 7. OF THE UNION OF SCIENTIFIC SOCIETIES.

This idea has occurred to several persons, as likely to lead to considerable advantages to science. If the various scientific societies could unite in the occupation of one large building, considerable economy would result from the union. By properly arranging their evenings of meeting, one meeting-room only need be required. The libraries might either be united, or arranged in adjoining rooms; and such a system would greatly facilitate the inquiries of scientific persons.

Whether it would be possible to reunite in any way the different societies to the Royal Society, might be a delicate question; but although, on some accounts, desirable, that event is not necessary for the purpose of their having a common residence.

The Medico-Botanical Society might, perhaps, from sympathy, be the first to which the Royal Society would apply; and by a proper interchange of diplomas, [A thing well understood by the INITIATED, both at HOME and ABROAD.] the two societies might be inoculated with each other. But even here some tact would be required; the Medico-Botanical is a little particular about the purity of its written documents, and lately attributed blame to one of its officers for some slight tampering with them, a degree of illiberality which the Council of the Royal Society are far from imitating.

The Geological and the Astronomical Societies nourish no feelings of resentment to the parent institution for their early persecution; and though they have no inducement to seek, would scarcely refuse any union which might be generally advantageous to science.

CONCLUSION.

In a work on the Decline of Science, at a period when England has so recently lost two of its brightest ornaments, I should hardly be excused if I omitted to devote a few words to the names of Wollaston and of Davy. Until the warm feelings of surviving kindred and admiring friends shall be cold as the grave from which remembrance vainly recalls their cherished forms, invested with all the life and energy of recent existence, the volumes of their biography must be sealed. Their contemporaries can expect only to read their eloge.

In habits of intercourse with both those distinguished individuals, sufficiently frequent to mark the curiously different structure of their minds, I was yet not on such terms even with him I most esteemed, as to view his great qualities through that medium which is rarely penetrated by the eyes of long and very intimate friendship.

Caution and precision were the predominant features of the character of Wollaston, and those who are disposed to reduce the number of principles, would perhaps justly trace the precision which adorned his philosophical, to the extreme caution which pervaded his moral character. It may indeed be questioned whether the latter quality will not in all persons of great abilities produce the former.

Ambition constituted a far larger ingredient in the character of Davy, and with the daring hand of genius he grasped even the remotest conclusions to which a theory led him. He seemed to think invention a more common attribute than it really is, and hastened, as soon as he was in possession of a new fact or a new principle, to communicate it to the world, doubtful perhaps lest he might not be anticipated; but, confident in his own powers, he was content to give to others a chance of reaping some part of that harvest, the largest portion of which he knew must still fall to his own share.

Dr. Wollaston, on the other hand, appreciated more truly the rarity of the inventive faculty; and, undeterred by the fear of being anticipated, when he had contrived a new instrument, or detected a new principle, he brought all the information that he could collect from others, or which arose from his own reflection, to bear upon it for years, before he delivered it to the world.

The most singular characteristic of Wollaston's mind was the plain and distinct line which separated what he knew from what he did not know; and this again, arising from his precision, might be traced to caution.

It would, however, have been visible to such an extent in few except himself, for there were very few so perfectly free from vanity and affectation. To this circumstance may be attributed a peculiarity of manner in the mode in which he communicated information to those who sought it from him, which was to many extremely disagreeable. He usually, by a few questions, ascertained precisely how much the inquirer knew upon the subject, or the exact point at which his ignorance commenced, a process not very agreeable to the vanity of mankind; taking up the subject at this point, he would then very clearly and shortly explain it.

His acquaintance with mathematics was very limited. Many years since, when I was an unsuccessful candidate for a professorship of mathematics, I applied to Dr. W. for a recommendation; he declined it, on the ground of its not being his pursuit. I told him I asked it, because I thought it would have weight, to which he replied, that it ought to have none whatever. There is no doubt his view was the just one. Yet such is the state of ignorance which exists on these subjects, that I have several times heard him mentioned as one of the greatest mathematicians of the age. [This of course could only have happened in England.] But in this as in all other points, the precision with which he comprehended and retained all he had ever learned, especially of the elementary applications of mathematics to physics, was such, that he possessed greater command over those subjects than many of far more extensive knowledge.

In associating with Wollaston, you perceived that the predominant principle was to avoid error; in the society of Davy, you saw that it was the desire to see and make known truth. Wollaston never could have been a poet; Davy might have been a great one.

A question which I put, successively, to each of these distinguished philosophers, will show how very differently a subject may be viewed by minds even of the highest order.

About the time Mr. Perkins was making his experiments on the compression of water, I was much struck with the mechanical means he had brought to bear on the subject, and was speculating on other applications of it, which I will presently mention.

Meeting Dr. Wollaston one morning in the shop of a bookseller, I proposed this question: If two volumes of hydrogen and one of oxygen are mixed together in a vessel, and if by mechanical pressure they can be so condensed as to become of the same specific gravity as water, will the gases under these circumstances unite and form water? "What do you think they will do?" said Dr. W. I replied, that I should rather expect they would unite. "I see no reason to suppose it," said he. I then inquired whether he thought the experiment worth making. He answered, that he did not, for that he should think it would certainly not succeed.

A few days after, I proposed the same question to Sir Humphry Davy. He at once said, "they will become water, of course;" and on my inquiring whether he thought the experiment worth making, he observed that it was a good experiment, but one which it was hardly necessary to make, as it must succeed.

These were off-hand answers, which it might perhaps be hardly fair to have recorded, had they been of persons of less eminent talent: and it adds to the curiosity of the circumstance to mention, that I believe Dr. Wollaston's reason for supposing no union would take place, arose from the nature of the electrical relations of the two gases remaining unchanged, an objection which did not weigh with the philosopher whose discoveries had given birth to it.

[The result of the experiment appeared, and still appears to me, to be of the highest importance; and I will shortly state the views with which it was connected. The next great discovery in chemistry to definite proportions, will be to find means of forming all the simple unions of one atom with one, with two, or with more of say other substance: and it occurred to me that the gaseous bodies presented the fairest chance of success; and that if wishing, for instance, to unite four atoms of one substance with one of another, we could, by mechanical means, reduce the mixed gases to the same specific gravity as the substance would possess which resulted from their union, then either that such union would actually take place, or the particles of the two substances would be most favourably situated for the action of caloric, electricity, or other causes, to produce the combination. It would indeed seem to follow, that if combination should take place under such circumstances, then the most probable proportion in which the atoms would unite, should be that which furnished a fluid of the least specific gravity: but until the experiments are made, it is by no means certain that other combinations might not be produced.]

The singular minuteness of the particles of bodies submitted by Dr. Wollaston to chemical analysis, has excited the admiration of all those who have had the good fortune to witness his experiments; and the methods he employed deserve to be much more widely known.

It appears to me that a great mistake exists on the subject. It has been adduced as one of those facts which prove the extraordinary acuteness of the bodily senses of the individual,—a circumstance which, if it were true, would add but little to his philosophical character; I am, however, inclined to view it in a far different light, and to see in it one of the natural results of the admirable precision of his knowledge.

During the many opportunities I have enjoyed of seeing his minute experiments, I remember but one instance in which I noticed any remarkable difference in the acuteness of his bodily faculties, either of his hearing, his sight, or of his sense of smell, from those of other persons who possessed them in a good degree. [This was at Mr. South's observatory, and the object was, the dots on the declination circle of his equatorial; but, in this instance, Dr. Wollaston did not attempt to TEACH ME HOW TO SEE THEM.]

He never showed me an almost microscopic wire, which was visible to his, and invisible to my own eye: even in the beautiful experiments he made relative to sounds inaudible to certain ears, he never produced a tone which was unheard by mine, although sensible to his ear; and I believe this will be found to have been the case by most of those whose minds had been much accustomed to experimental inquiries, and who possessed their faculties unimpaired by illness or by age.

It was a much more valuable property on which the success of such inquiries depended. It arose from the perfect attention which he could command, and the minute precision with which he examined every object. A striking illustration of the fact that an object is frequently not seen, FROM NOT KNOWING HOW TO SEE IT, rather than from any defect in the organ of vision, occurred to me some years since, when on a visit at Slough. Conversing with Mr. Herschel on the dark lines seen in the solar spectrum by Fraunhofer, he inquired whether I had seen them; and on my replying in the negative, and expressing a great desire to see them, he mentioned the extreme difficulty he had had, even with Fraunhofer's description in his hand and the long time which it had cost him in detecting them. My friend then added, "I will prepare the apparatus, and put you in such a position that they shall be visible, and yet you shall look for them and not find them: after which, while you remain in the same position, I will instruct you how to see them, and you shall see them, and not merely wonder you did not see them before, but you shall find it impossible to look at the spectrum without seeing them."

On looking as I was directed, notwithstanding the previous warning, I did not see them; and after some time I inquired how they might be seen, when the prediction of Mr. Herschel was completely fulfilled.

It was this attention to minute phenomena which Dr. Wollaston applied with such powerful effect to chemistry. In the ordinary cases of precipitation the cloudiness is visible in a single drop as well as in a gallon of a solution; and in those cases where the cloudiness is so slight, as to require a mass of fluid to render it visible, previous evaporation, quickly performed on slips of window glass, rendered the solution more concentrated.

The true value of this minute chemistry arises from its cheapness and the extreme rapidity with which it can be accomplished: it may, in hands like those of Wollaston, be used for discovery, but not for measure. I have thought it more necessary to place this subject on what I consider its true grounds, for two reasons. In the first place, I feel that injustice has been done to a distinguished philosopher in attributing to some of his bodily senses that excellence which I think is proved to have depended on the admirable training of his intellectual faculties. And, in the next place, if I have established the fact, whilst it affords us better means of judging of such observations as lay claim to an accuracy "MORE THAN HUMAN," it also opens, to the patient inquirer into truth, a path by which he may acquire powers that he would otherwise have thought were only the gift of nature to a favoured few.

APPENDIX, No. 1.

In presenting to my readers the account of the meeting of men of science at Berlin, in the autumn of 1828, I am happy to be able to state, that its influence has been most beneficial, and that the annual meeting to be held in 1831, will take place at Vienna, the Emperor of Austria having expressed a wish that every facility which his capital affords should be given to promote its objects.

It is gratifying to find that a country, which has hitherto been considered adverse to the progress of knowledge, should become convinced of its value; and it is sincerely to be hoped, that every one of the numerous members of the Society will show, by his conduct, that the paths of science are less likely than any others to interfere with those of politics.

ACCOUNT OF THE GREAT CONGRESS OF PHILOSOPHERS AT BERLIN, ON THE 18TH OF SEPTEMBER 1828. FROM THE EDINBURGH JOURNAL OF SCIENCE, APRIL, 1829.

The existence of a large society of cultivators of the natural sciences meeting annually at some great capital, or some central town of Europe, is a circumstance almost unknown to us, and deserving of our attention, from the important advantages which may arise from it.

About eight years ago, Dr. Okens, of Munich, suggested a plan for an annual meeting of all Germans who cultivated the sciences of medicine and botany. The first meeting, of about forty members, took place at Leipsic, in 1822, and it was successively held at Halle, Wurtzburg, Frankfort on the Maine, Dresden, Munich, and Berlin. All those who had printed a certain number of sheets of their inquiries on these subjects were considered members of this academy.

The great advantages which resulted to these sciences from the communication of observations from all quarters of Germany, soon induced an extension of the plan, and other departments of natural knowledge were admitted, until, at the last meeting, the cultivators even of pure mathematics were found amongst the ranks of this academy.

Several circumstances, independent of the form and constitution of the academy, contributed to give unwonted splendour to the last meeting, which took place at Berlin in the middle of September of the last year.

The capital selected for its temporary residence is scarcely surpassed by any in Europe in the number and celebrity of its savans.

The taste for knowledge possessed by the reigning family, has made knowledge itself fashionable; and the severe sufferings of the Prussians previous to the war, by which themselves and Europe were freed, have impressed on them so strongly the lesson that "knowledge is power," that its effects are visible in every department of the government; and there is no country in Europe in which talents and genius so surely open for their possessors the road to wealth and distinction.

Another circumstance also contributed its portion to increase the numbers of the meeting of the past year. The office of president, which is annually changed, was assigned to M. Alexander de Humboldt. The universality of his acquirements, which have left no branch within the wide range of science indifferent or unexplored, has connected him by friendship with almost all the most celebrated philosophers of the age; whilst the polished amenity of his manners, and that intense desire of acquiring and of spreading knowledge, which so peculiarly characterizes his mind, renders him accessible to all strangers, and insures for them the assistance of his counsel in their scientific pursuits, and the advantage of being made known to all those who are interested or occupied in similar inquiries.

Professor Lichtenstein, (Director of the Museum of Zoology,) as secretary of the academy, was indefatigable in his attentions, and most ably seconded the wishes of its distinguished president.

These two gentlemen, assisted by several of the residents at Berlin, undertook the numerous preliminary arrangements necessary for the accommodation of the meeting.

On the 18th of September, 1828, there were assembled at Berlin 377 members of the academy, whose names and residences (in Berlin) were printed in a small pamphlet, and to each name was attached a number, to indicate his seat in the great concert room, in which the morning meetings took place. Each member was also provided with an engraved card of the hall of meeting, on which the numbers of the seats were printed in black ink, and his own peculiar seat marked in red ink, so that every person immediately found his own place, and knew where to look for any friend whom he might wish to find.

At the hour appointed for the opening of the meeting, the members being assembled, and the galleries and orchestra being filled by an assemblage of a large part of the rank and beauty of the capital, and the side-boxes being occupied by several branches of the royal family, and by the foreign ambassadors, the session of the academy was opened by the eloquent address of the president.

SPEECH made at the Opening of the Society of German Naturalists and Natural Philosophers at Berlin, the 18th of September, 1828.—By ALEXANDER VON HUMBOLDT.

Since through your choice, which does me so much honour, I am permitted to open this meeting, the first duty which I have to discharge is one of gratitude. The distinction which has been conferred on him who has never yet been able to attend your excellent society, is not the reward of scientific efforts, or of feeble and persevering attempts to discover new phenomena, or to draw the light of knowledge from the unexplored depths of nature. A finer feeling, however, directed your attention to me. You have assured me, that while, during an absence of many years, and in a distant quarter of the globe, I was labouring in the same cause with yourselves, I was not a stranger in your thoughts. You have likewise greeted my return home, that, by the sacred tie of gratitude, you might bind me still longer and closer to our common country.

What, however, can the picture of this, our native land, present more agreeable to the mind, than the assembly which we receive to-day for the first time within our walls; from the banks of the Neckar, the birth-place of Kepler and of Schiller, to the remotest border of the Baltic plains; from hence to the mouths of the Rhine, where, under the beneficent influence of commerce, the treasuries of exotic nature have for centuries been collected and investigated, the friends of nature, inspired with the same zeal, and, urged by the same passion, flock together to this assembly. Everywhere, where the German language is used, and its peculiar structure affects the spirit and disposition of the people. From the Great European Alps, to the other side of the Weichsel, where, in the country of Copernicus, astronomy rose to renewed splendour; everywhere in the extensive dominions of the German nation we attempt to discover the secret operations of nature, whether in the heavens, or in the deepest problems of mechanics, or in the interior of the earth, or in the finely woven tissues of organic structure.

Protected by noble princes, this assembly has annually increased in interest and extent. Every distinction which difference of religion or form of government can occasion is here annulled. Germany manifests itself as it were in its intellectual unity; and since knowledge of truth and performance of duty are the highest object of morality, that feeling of unity weakens none of the bonds which the religion, constitution, and laws of our country, have rendered dear to each of us. Even this emulation in mental struggles has called forth (as the glorious history of our country tells us,) the fairest blossoms of humanity, science, and art.

The assembly of German naturalists and natural philosophers since its last meeting, when it was so hospitably received at Munich, has, through the flattering interest of neighbouring states and academies, shone with peculiar lustre. Allied nations have renewed the ancient alliance between Germany and the ancient Scandinavian North.

Such an interest deserves acknowledgment the more, because it unexpectedly increases the mass of facts and opinions which are here brought into one common and useful union. It also recalls lofty recollections into the mind of the naturalist. Scarcely half a century has elapsed since Linne appears, in the boldness of the undertakings which he has attempted and accomplished, as one of the greatest men of the last century. His glory, however bright, has not rendered Europe blind to the merits of Scheele and Bergman. The catalogue of these great names is not completed; but lest I shall offend noble modesty, I dare not speak of the light which is still flowing in richest profusion from the North, nor mention the discoveries in the chemical nature of substances, in the numerical relation of their elements, or the eddying streams of electro-magnetic powers. [The philosophers here referred to are Berzelius and Oersted.] May those excellent persons, who, deterred neither by perils of sea or land, have hastened to our meeting from Sweden, Norway, Denmark, Holland, England, and Poland, point our the way to other strangers in succeeding years, so that by turns every part of Germany may enjoy the effects of scientific communication with the different nations of Europe.

But although I must restrain the expression of my personal feelings in presence of this assembly, I must be permitted at least to name the patriarchs of our national glory, who are detained from us by a regard for those lives so dear to their country;— Goethe, whom the great creations of poetical fancy have not prevented from penetrating the ARCANA of nature, and who now in rural solitude mourns for his princely friend, as Germany for one of her greatest ornaments;—Olbers, who has discovered two bodies where he had already predicted they were to be found;—the greatest anatomists of our age— Soemmering, who, with equal zeal, has investigated the wonders of organic structure, and the spots and FACULAE of the sun, (condensations and openings of the photosphere;) Blumenbach, whose pupil I have the honour to be, who, by his works and his immortal eloquence, has inspired everywhere a love of comparative anatomy, physiology, and the general history of nature, and who has laboured diligently for half a century. How could I resist the temptation to adorn my discourse with names which posterity will repeat, as we are not favoured with their presence?

These observations on the literary wealth of our native country, and the progressive developement of our institution, lead us naturally to the obstructions which will arise from the increasing number of our fellow-labourers, The chief object of this assembly does not consist, as in other societies whose sphere is more limited, in the mutual interchange of treatises, or in innumerable memoirs, destined to be printed in some general collection. The principal object of this Society is, to bring those personally together who are engaged in the same field of science. It is the immediate, and therefore more obvious interchange of ideas, whether they present themselves as facts, opinions, or doubts. It is the foundation of friendly connexion which throws light on science, adds cheerfulness to life, and gives patience and amenity to the manners.

In the most flourishing period of ancient Greece, the distinction between words and writing first manifested itself most strongly amongst a race, which had raised itself to the most splendid intellectual superiority, and to whose latest descendants, as preserved from the shipwreck of nations, we still consecrate our most anxious wishes. It was not the difficulty of interchange of ideas alone, nor the want of German science, which has spread thought as on wings through the world, and insured it a long continuance, that then induced the friends of philosophy and natural history in Magna Graecia and Asia Minor to wander on long journeys. That ancient race knew the inspiring influence of conversation as it extemporaneously, freely, and prudently penetrates the tissue of scientific opinions and doubts. The discovery of the truth without difference of opinion is unattainable, because the truth, in its greatest extent, can never be recognized by all, and at the same time. Each step, which seems to bring the explorer of nature nearer to his object, only carries him to the threshold of new labyrinths. The mass of doubt does not diminish, but spreads like a moving cloud over other and new fields; and whoever has called that a golden period, when difference of opinions, or, as some are accustomed to express it, the disputes of the learned, will be finished, has as imperfect a conception of the wants of science, and of its continued advancement, as a person who expects that the same opinions in geognosy, chemistry, or physiology, will be maintained for several centuries.

The founders of this society, with a deep sense of the unity of nature, have combined in the completest manner, all the branches of physical knowledge, and the historical, geometrical, and experimental philosophy. The names of natural historian and natural philosopher are here, therefore, nearly synonymous, chained by a terrestrial link to the type of the lower animals. Man completes the scale of higher organization. In his physiological and pathological qualities, he scarcely presents to us a distinct class of beings. As to what has brought him to this exalted object of physical study, and has raised him to general scientific investigation, belongs principally to this society. Important as it is not to break that link which embraces equally the investigation of organic and inorganic nature, still the increasing ties and daily developement of this institution renders it necessary, besides the general meeting which is destined for these halls, to have specific meetings for single branches of science. For it is only in such contracted circles,—it is only among men whom reciprocity of studies has brought together, that verbal discussions can take place. Without this sort of communication, would the voluntary association of men in search of truth be deprived of an inspiring principle.

Among the preparations which are made in this city for the advancement of the society, attention has been principally paid to the possibility of such a subdivision into sections. The hope that these preparations will meet with your approbation, imposes upon me the duty of reminding you, that, although you had entrusted to two travellers, equally, the duty of making these arrangements, yet it is to one alone, my noble friend, M. Lichtenstein, that the merit of careful precaution and indefatigable activity is due. Out of respect to the scientific spirit which animates the Society of German Naturalists and Natural Philosophy, and in acknowledgment of the utility of their efforts, government have seconded all our wishes with the greatest cheerfulness.

In the vicinity of the place of meeting, which has in this manner been prepared for our general and special labours, are situated the museums dedicated to anatomy, zoology, oryctognosy, and geology.

They exhibit to the naturalist a rich mine for observation and critical discussion. The greater number of these well-arranged collections have existed, like the University of Berlin, scarcely twenty years. The oldest of them, to which the Botanical Garden, (one of the richest in Europe) belongs, have during this period not only been increased, but entirely remodelled. The amusement and instruction derived from such institutions, call to our minds, with deep feelings of gratitude, that they are the work of that great monarch, who modestly and in simple grandeur, adorns every year this royal city with new treasures of nature and art; and what is of still greater value than the treasures themselves,— what inspires every Prussian with youthful strength, and with an enthusiastic love for the ancient reigning family,—that he graciously attaches to himself every species of talent, and extends with confidence his royal protection to the free cultivation of the understanding.

This was followed by a paper on magnetism, by Professor Oersted; and several other memoirs were then read.

The arrival of so many persons of similar pursuit, (for 464 members were present,) rendered it convenient to have some ordinary, at which those who chose might dine, and introduce their friends or families. This had been foreseen, and his Majesty had condescended to allow the immense building used for the exercise of his troops, to be employed for this purpose. One-third of it was floored on the occasion, and tables were arranged, at which, on one occasion, 850 persons sat down to dinner. On the evening of the first day, M. de Humboldt gave a large SOIREE in the concert rooms attached to the theatre. About 1200 persons assembled on this occasion, and his Majesty the King of Prussia honoured with his presence the fete of his illustrious chamberlain. The nobility of the country, foreign princes, and foreign ambassadors, were present. It was gratifying to observe the princes of the blood mingling with the cultivators of science, and to see the heir-apparent to the throne, during the course of the evening, engaged in conversation with those most celebrated for their talents, of his own, or of other countries.

Nor were the minor arrangements of the evening beneath the consideration of the President. The words of the music selected for the concert, were printed and distributed to the visitors. The names of the most illustrious philosophers which Germany had produced, were inscribed in letters of gold at the end of the great concert room.

In the first rank amongst these stood a name which, England, too, enrolls amongst the brightest in her scientific annals; and proud, as well she may be, of having fostered and brought to maturity the genius of the first Herschel, she has reaped an ample reward in being able to claim as entirely her own, the inheritor of his talents and his name.

The six succeeding days were occupied, in the morning, by a meeting of the academy, at which papers of general interest were read. In the afternoon, through the arrangement of M. de Humboldt and M. Lichtenstein, various rooms were appropriated for different sections of the academy. In one, the chemical philosophers attended to some chemical memoir, whilst the botanists assembled in another room, the physiologists in a third, and the natural philosophers in a fourth. Each attended to the reading of papers connected with their several sciences. Thus every member was at liberty to choose that section in which he felt most interest at the moment, and he had at all times power of access to the others. The evenings were generally spent at some of the SOIREES of the savans, resident at Berlin, whose hospitality and attentions to their learned brethren of other countries were unbounded. During the unoccupied hours of the morning, the collections of natural history, which are rapidly rising into importance, were open to examination; and the various professors and directors who assisted the stranger in his inquiries, left him equally gratified by the knowledge and urbanity of those who so kindly aided him.

A map of Europe was printed, on which those towns only appeared which had sent representatives to this scientific congress; and the numbers sent by different kingdoms appeared by the following table, which was attached to it;—

Russia......... 1
Austria........ 0
England........ 1
Holland........ 2
Denmark........ 7
France 1
Sardinia 0
Prussia........ 95
Bavaria........ 12
Hanover........ 5
Saxony 21
Wirtemburg 2
Sweden 13
Naples 1
Poland 3
German States..... 43

——-   206

Berlin 172

——-   378

The proportion in which the cultivators of different sciences appeared, was not easy to ascertain, because there were few amongst the more eminent who had not added to more than one branch of human knowledge. The following table, though not professing to be very accurate, will afford, perhaps, a tolerably fair view:—

Geometers............. 11
Astronomers........... 5
Natural Philosophers . 23
                — 39

Mines............ 5
Mineralogy 16

Geology.......... 9
                — 30

Chemistry........... 18
Geography........... 8
Anatomy............. 12

Zoology............ 14
Natural History.... 8
Botany.............. 35
                — 57

Physicians....... 175
Amateurs 9
Various 35
                —-381

A medal was struck in commemoration of this meeting, and it was proposed that it should form the first of a series, which should comprise all those persons most celebrated for their scientific discoveries in the past and present age.

APPENDIX, No. 2.

An examination into some charges brought against one of the twenty-four candidates, mentioned in a note as having their names suspended in the meeting-room of the Royal Society, at one time, has caused a printed pamphlet to be circulated amongst the members of the Society. Of the charges themselves I shall offer no opinion, but entreat every member to judge for himself. I shall, however, make one extract, which tends to show how the ranks of the Society are recruited.

EXTRACT FROM A PRINTED LETTER FROM A. F. M. TO J. G. CHILDREN, ESQ. DATED, 22, UPPER BEDFORD-PLACE, MARCH 13, 1830.

"When I wished you to Propose me at the Geological Society, you asked me why you should not propose me also at the Royal Society; and my answer was, that it was an honour to which I did not think I could aspire; that my talents were too insignificant to warrant such pretensions. Many days passed, and still you pressed me on the subject, because your partiality made you think me deserving of the honour; but I resisted, really through modesty, not that I did not covet the distinction, until something was said of my paper on the meteoric mass of iron of Brazil, which was published some years ago in the Transactions of the Royal Society; when you insisted on proposing me, and I assented gratefully, because I was and am desirous of being a Fellow of the Royal Society, if I can be supposed worthy of having my name so honourably enrolled."

EXTRACT FROM A LETTER OF J. G. CHILDREN, ESQ. TO A. F. M. ESQ. DATED, BRITISH MUSEUM, MARCH 24, 1830.

"All that you have said respecting your being a candidate for admission into the Royal Society, is correct to the letter. I pressed the subject upon you, and I would do it again to-morrow, were it necessary."

Here, then, we find Mr. Children, who has been on the Council of the Royal Society, and who was, a few years since, one of its Secretaries, pressing one of his friends to become, and actually insisting on proposing him as, a Fellow of the Royal Society, He must have been well aware of the feelings which prevail amongst the Council as to the propriety of such a step, and by publishing the fact, seems quite satisfied that such a course is advantageous to the interests of the Society. That similar applications were not unfrequently made in private, is well known; but it remains for the Society to consider whether, now they are publicly and officially announced to them, it will sanction this mode of augmenting the already numerous list of its fellows.

APPENDIX, No. 3,

LIST OF THE MEMBERS OF THE ROYAL SOCIETY, WHO HAVE CONTRIBUTED TO THE

PHILOSOPHICAL TRANSACTIONS, OR HAVE BEEN ON THE COUNCIL.

N. B.—The Numbers are made up to the present year for the Papers, but only to 1827 for Members of the Council.

No. of Papers printed in Phil. Trans.	No. of years on Council.	
3		Aberdeen, Earl of.
3	3	Abernethy, John.
	2	Allan, Thomas.
3		Allen, William.
	1	Arden, Lord.
	1	Atholl, Duke of.
7	2	Babbage, Charles,
	1	Babington, William.
1	2	Baily,Francis.
9		Barlow, Peter. (C)
	2	Barnard, Sir F. Augusta.
	5	Barrow, John.
2		Bauer, Francis.
1		Bayley, John.

	1	Beaufort, Francis.
	2	Beaufoy, Henry.
5		Bell, Charles.
	1	Bingley, Robert.
	1	Blackburne, John.
	3	Blake, William.
1	3	Blane, Sir Gilbert.
1	1	Blizard, Sir William.
1	1	Bostock, John.
12	10	Brande, Wm. Thos. (C)
16		Brewster, David. (C)
6	1	Brodie, B. Collins. (C)
1		Bromhead Sir E. F.
3		Brougham, Henry.
	1	Browne, Henry.
	1	Brown, Robert.
	2	Brownlow, Earl.
1		Buckland, Rev. W. (C)
	1	Burney, Rev. C. Parr.
	1	Canterbury, Archbp. of.
	1	Carew, Rt. Hon. R. P.
7		Carlisle, Sir Anthony.
	2	Carlisle, Nicholas.
1		Carne, Joseph.
	1	Carrington, Sir C. E.
	2	Charleville, Earl of.
7	2	Chenevix, Richard. (C)
3	4	Children, John George.
10		Christie, Sam. Hunter.
	1	Clerk, Sir George.
2		Clift, William.
9		Cloyne, Bishop of. (C)
	2	Colby, Colonel Thomas.

	1	Colebrooke, Henry T.
2	2	Cooper, Sir Astley P. (C)
	1	Crichton, Sir Alex.
	5	Croker, John Wilson.
	1	Cullum, Sir T. Gery.
2		Dalton, John.
	2	Darnley, Earl of
1		Darwin, Robert Waring.
1		Davis, John Francis.
2		Davy, Edmund.
13		Davy, John.
3		Dyllwin, Lewis Weston.
1		Dollond, George.
	1	Dudley and Ward, Visc.
2		Earle, Henry.
	1	Egremont, Earl of.
1		Fallows, Rev. Fearon.
8		Faraday, Michael.
	1	Farnborough, Lord.
1		Fisher, Rev. George.
	1	Fly, Rev. Henry.
2		Foster, Henry.
1	1	Frankland, Sir Thomas.
1		Gibbes, Sir Geo, Smith.
2	13	Gilbert, Davies.
	2	Gillies, John.
5		Goldingham, John.
3	1	Gompertz, Benjamin.
	1	Goodenough, George T.
	2	Gordon, Sir James W.
3		Granville, Augustus B.
1		Greatorex, Thomas.
1		Greenough, Geo.Bellas.

1		Griffiths, John.
3	1	Groombridge, Stephen.
	1	Halford, Sir Henry.
2		Hall, Basil.
	1	Hamilton, Wm. Rich.
	2	Hardwicke, Earl of.
2		Harvey, George.
1		Harwood, J.
16	10	Hatchett, Charles. (C)
	1	Hawkins, John.
2	2	Heberden, William.
9		Hellins, Rev. John, (C)
	1	Henley, Morton Lord.
10		Henry, William. (C)
12	6	Herschel, John F.W. (C)
	1	Hoare, Henry Hugh
	1	Hoare, Sir Richard Colt.
	2	Hobhouse, Sir Benj.
1		Holland, Henry.
109	16	Home, Sir Everard. (C)
2		Hope, Thomas Charles.
1		Hosack, David.
1	1	Horsburgh, James.
1		Howard, Luke.
2		Hume, Sir Abraham.
7	2	Ivory, James.C.
	1	Jekyll, Joseph.
4	1	Johnson, Jas. Rawlins.
13	7	Kater, Capt. Henry. (C)
2		Kidd, John.
24	1	Knight, Thomas A. (C)
1	1	Konig, Charles.
	2	Lambert, Aylmer B.

	1	Lansdowne, Marquis of.
1	1	Latham, John.
2		Lax, Rev. William.
1		Leach, William Elford.
	1	Lowther, Viscount.
2		Macartney, James.
2		Macdonald, Lieut. Col.
	1	Mac Grigor, Sir James.
	2	Mac Leay, Alexander.
	1	Mansfield, Earl of
4	11	Marsden, William.
	1	Mathias, Thomas Jas.
	3	Maton, William George.
1		Miller, Lieut. Col. G.
	2	Montagu, Matthew.
7	4	Morgan, William.
	1	Mount Edgecumbe, Earl of.
	3	Murdoch, Thomas.
	2	Nicholl, Rt. Hon. Sir J.
	1	Norfolk, Duke of.
	2	Ord, Craven.
1		Parry, Charles Henry.
	1	Pepys, Sir Lucas.
6	2	Pepys, Wm. Hasledine.
7		Philip, A. P. Wilson.
1		Phillips, Richard.
	2	Pitt, William Morton.
1	29	Planta, Joseph.
19	17	Pond, John. (C)
2		Powell, Rev. Baden.
2		Prinsep, James.
4	1	Prout William.
	1	Rackett, Rev. Thomas.

	1	Redesdale, Lord.
	2	Reeves, John.
5	3	Rennell, James (C)
1		Rennie, George.
4		Ritchie,
1		Robertson, James.
	1	Rogers, Samuel.
2	1	Roget, Peter Mark.
	3	Rudge, Edward.
12		Sabine, Edward. (C)
	2	Sabine, Joseph.
	1	St. Aubyn, Sir John.
3		Scoresby, jun. William.
2		Scott, John Corse.
3	1	Seppings, Sir Robert. (C)
1		Sewell, Sir John.
	3	Somerset, Duke of.
	3	Sotheby, William.
3	2	South, James. (C)
	5	Spencer, Earl.
	3	Stanley, Sir John Thos.
	3	Staunton, Sir Geo. Thos.
	2	Stowell,Lord.
	1	Sumner, George Holme.
1		Thomas, Honoratus L.
2		Thomson, Thomas.
1		Tiarks, Dr. John Lewis.
1		Troughton, Edward. (C)
2		Ure, Andrew.
	2	Warburton, Henry.
1		Weaver, Thomas.
1		Whewell, William.
3		Whidbey, Joseph.

2	3	Wilkins, Charles.
3		Williams, John Lloyd.
1	1	Wilson, Sir Giffin.
	2	Wilson, Gloucester.
	1	Yorke, Rt. Hon. Chas.

I had intended to have printed a list of those persons to whom the Royal Society had in past years awarded the Copley medals, and the reasons for which they were given; but having applied to the Council for permission to employ an amanuensis, to copy those awards, either from the minutes, or from the volumes of the Philosophical Transactions, I was surprised at receiving a refusal. I confess it appeared to me, that as a whole, those adjudications did us credit, although I doubted the propriety of many individual cases. As, however, the Council seem to have had a different opinion, and as I had made the application through courtesy, I shall decline printing a list, every individual portion of which has been already published in many ways, although the whole has never been printed in a collected form.

*** END OF THIS BOOK "DECLINE OF SCIENCE IN ENGLAND" ***

www.ingramcontent.com/pod-product-compliance
Lightning Source LLC
Chambersburg PA
CBHW051644170526
45167CB00001B/322